Inverse Obstacle Scattering with Non-Over-Determined Scattering Data

Alexander G. Ramm

Synthesis Lectures on Mathematics and Statistics

Editor
Steven G. Krantz, *Washington University, St. Louis*

Inverse Obstacle Scattering with Non-Over-Determined Scattering Data
Alexander G. Ramm
2019

Aspects of Differential Geometry IV
Esteban Calviño-Louzao, Eduardo García-Río, Peter Gilkey, JeongI Iyeong Park, and Ramón Vázquez-Lorenzo
2019

Symmetry Problems. The Navier–Stokes Problem.
Alexander G. Ramm
2019

PDE Models for Atherosclerosis Computer Implementation in R
William E. Schiesser
2018

An Introduction to Partial Differential Equations
Daniel J. Arrigo
2017

Numerical Integration of Space Fractional Partial Differential Equations: Vol 2 – Applicatons from Classical Integer PDEs
Younes Salehi and William E. Schiesser
2017

Numerical Integration of Space Fractional Partial Differential Equations: Vol 1 – Introduction to Algorithms and Computer Coding in R
Younes Salehi and William E. Schiesser
2017

Inverse Obstacle Scattering with Non-Over-Determined Scattering Data
Alexander G. Ramm

ISBN: 978-3-031-01290-7   paperback
ISBN: 978-3-031-02418-4   ebook
ISBN: 978-3-031-00264-9   hardcover

DOI 10.1007/978-3-031-02418-4

A Publication in the Springer series series
SYNTHESIS LECTURES ON MATHEMATICS AND STATISTICS

Lecture #27
Series Editor: Steven G. Krantz, *Washington University, St. Louis*
Series ISSN
Print 1938-1743   Electronic 1938-1751

Inverse Obstacle Scattering with Non-Over-Determined Scattering Data

Alexander G. Ramm
ramm@ksu.edu

SYNTHESIS LECTURES ON MATHEMATICS AND STATISTICS #27

ABSTRACT

The inverse obstacle scattering problem consists of finding the unknown surface of a body (obstacle) from the scattering $A(\beta; \alpha; k)$, where $A(\beta; \alpha; k)$ is the scattering amplitude, $\beta; \alpha \in S^2$ is the direction of the scattered, incident wave, respectively, S^2 is the unit sphere in the $\mathbb{R}^3$ and $k > 0$ is the modulus of the wave vector. The scattering data is called non-over-determined if its dimensionality is the same as the one of the unknown object. By the dimensionality one understands the minimal number of variables of a function describing the data or an object. In an inverse obstacle scattering problem this number is 2, and an example of non-over-determined data is $A(\beta) := A(\beta; \alpha_0; k_0)$. By sub-index 0 a fixed value of a variable is denoted.

It is proved in this book that the data $A(\beta)$, known for all β in an open subset of S^2, determines uniquely the surface S and the boundary condition on S. This condition can be the Dirichlet, or the Neumann, or the impedance type.

The above uniqueness theorem is of principal importance because the non-over-determined data are the minimal data determining uniquely the unknown S. There were no such results in the literature, therefore the need for this book arose. This book contains a self-contained proof of the existence and uniqueness of the scattering solution for rough surfaces.

KEYWORDS

direct scattering, inverse obstacle scattering, numerical analysis, mathematical analysis, non-over-determined data

To Luba

Contents

Preface

The aim of this book is to present the theory, developed by the author. The main result is the proof of the uniqueness of the solution to inverse obstacle scattering problem *with non-over-determined scattering data*.

By the non-over-determined scattering data, the value of $A(\beta) := A(\beta, \alpha_0, k_0)$ is meant for all $\beta \in S_0^2$, where S_0^2 is an open subset of the unit sphere S^2, and a fixed values of $\alpha = \alpha_0 \in S^2$ and $k = k_0 > 0$. Here and throughout S^2 is the unit sphere in $\mathbb{R}^3$. By $A(\beta, \alpha, k)$ the scattering amplitude is denoted, α and β are the unit vectors in the directions of the incident wave and the scattered wave, respectively, $\alpha_0 \in S^2$ is an arbitrary fixed direction, and $k_0 > 0$ is an arbitrary fixed value of k.

The unknown object in an inverse obstacle scattering problem is the surface S of an obstacle, and the obstacle is a bounded domain D in $\mathbb{R}^3$. A surface S in $\mathbb{R}^3$ is described by a function of two variables. So, the scattering data, $A(\beta)$, and the unknown object, S, depend on the same number of variables.

In this case the inverse obstacle scattering problem is the problem with *non-over-determined scattering data $A(\beta)$*.

The uniqueness theorem for the solution of an inverse scattering problem with non-over-determined data is of principal importance because these data are minimal data needed for the unique identification of the unknown object, which is the unknown surface S of the obstacle in this book (and the unknown compactly supported potential in the author's monograph [9]). These data are necessary and sufficient for the unique identification of S.

The obstacle inverse scattering problem consists of finding the surface S of the obstacle D and the boundary condition on S from the scattering data. The scattering data are the values of the scattering amplitude $A(\beta, \alpha, k)$, where $\beta, \alpha \in S^2$ and $k \in [0, \infty)$. So, the scattering amplitude depends, in general, on five variables.

By α_0 (or k_0) a fixed value of α (or k) is denoted. Thus, $A(\beta) := A(\beta, \alpha_0, k_0)$ is a function of two variables.

The scattering amplitude is defined by the asymptotic formula:

$$u(x, \alpha, k) = u_0(x, \alpha, k) + \frac{e^{ikr}}{r} A(\beta, \alpha, k) + o\left(\frac{1}{r}\right), \quad r \to \infty, \quad \beta := \frac{x}{r}, \tag{1}$$

where $u_0 = e^{ik\alpha \cdot x}$ is the incident plane wave, $\alpha \cdot x$ is the dot product of two vectors, and $u(x, \alpha, k)$ is the scattering solution.

The scattering solution is the unique solution to the scattering problem:

$$(\nabla^2 + k^2)\, u \;=\; 0 \quad \text{in} \quad D' := \mathbb{R}^3 \setminus D, \tag{2}$$

$$\Gamma u|_S \;=\; 0, \tag{3}$$

$$u \;=\; u_0 + v, \tag{4}$$

$$\frac{\partial v}{\partial r} - ikv \;=\; o\!\left(\frac{1}{r}\right), \quad r \to \infty. \tag{5}$$

Here $k = \text{const} > 0$, v is the scattered field, by Γu we mean the boundary condition, and (5) is the radiation condition. We assume that the boundary condition is either the Dirichlet one, $\Gamma u = u$, or the Neumann one, $\Gamma u = u_N$, or the impedance one

$$\Gamma u = (u_N + \zeta(s)u)\,|_S = 0. \tag{6}$$

We also assume that $D \subset \mathbb{R}^3$ is a bounded domain with a C^2-smooth boundary S, N is the unit outer normal on S, D' is the complement of D, $D' = \mathbb{R}^3 \setminus D$, the boundary impedance $\zeta(s)$ is a continuous function on S,

$$\text{Im}\,\zeta(s) \geq 0. \tag{7}$$

Condition (7) guarantees uniqueness of the scattering solution $u(x, \alpha, k)$.

In Appendix A the smoothness requirements on the boundary S are reduced very much and are close to the minimal possible.

The history of the inverse obstacle scattering is not very long. In the beginning of the 1960s. M. Schiffer proved that the data $A(\beta, \alpha_0, k)$ known for all $\beta \in S^2$ and all $k > 0$, determine uniquely S for the Dirichlet boundary condition. The data $A(\beta, \alpha_0, k)$ is a function of three variables while S is described by a function of two variables. Therefore the data in this inverse problem is over-determined, that is, the dimensionality of the data is larger than that of the unknown object.

In 1985, the author proved that the surface S and the boundary condition on S are uniquely determined by the fixed energy scattering data $A(\beta, \alpha, k_0)$, known for all β on an open subset of S^2, all β on an open subset of S^2 and a fixed $k_0 > 0$; see [9]. These data allow one to find uniquely S, the boundary condition on S, and the function $\zeta(s)$ on S, where $\zeta(s)$ is the boundary impedance, $\text{Im}\,\zeta(s) \geq 0$.

But the fixed-energy data $A(\beta, \alpha, k_0)$ are over-determined. The function $A(\beta, \alpha, k_0)$ depends on four variables since a unit vector is defined by two variables.

The first uniqueness theorem for the solution of the obstacle inverse problem with non-over-determined data was proved by the author, see papers [11], [12], where the obstacles were assumed strictly convex, paper [14], where the convexity assumption was dropped, and the monographs [9], [10]. A proof of this theorem is presented in this book.

The first results on inverse potential scattering problem with non-over-determined data were obtained in [13], [14], [22], [26] and in the monographs [10], [9]. For the inverse potential scattering the first and the only results were obtained by the author in [19], [16], [17], and in the monograph [9].

Also, a numerical method for solving the inverse obstacle scattering data with non-over-determined data is presented in this book. Critique of the published numerical methods for solving this problem is given in [18], pp. 247–253.

For potential inverse scattering with non-over-determined data a numerical method is developed in [15] and [23].

To make this presentation self-contained the author gives all the proofs of the basic results in detail. The aim of this book is to present the theory, developed by the author. We prove the uniqueness of the solution to inverse obstacle scattering problem *with non-over-determined scattering data.*

Alexander G. Ramm
June 2019

CHAPTER 1

Introduction

Let $D \subset \mathbb{R}^3$ be a bounded domain with a sufficiently smooth boundary S. Consider the direct scattering problem for an obstacle D. This problem consists of finding the scattering solution $u(x, \alpha, k)$ as the solution to the Helmholtz equation

$$
\begin{aligned}
\left(\nabla^2 + k^2\right) u &= 0 \quad \text{in } D', \quad k = \text{const} > 0, & (1.1) \\
u &= u_0 + v, & (1.2) \\
\frac{\partial v}{\partial N} + \zeta(s) v &= 0 \quad \text{on } S, \quad \operatorname{Im} \zeta(s) \geq 0, & (1.3) \\
\frac{\partial v}{\partial r} - i k v &= o\left(\frac{1}{r}\right), \quad r := |x| \to \infty. & (1.4) \\
u_0 &= e^{ik\alpha \cdot x}, \quad \alpha \in S^2, & (1.5) \\
\operatorname{Im}\zeta(s) &\geq 0, & (1.6)
\end{aligned}
$$

where $N(s)$ is the unit normal to S pointing into $D' = \mathbb{R}^3 \setminus D$, $\zeta(s)$ is a continuous function, S is C^2-smooth, and the $o\left(\frac{1}{r}\right)$ in (1.4) is uniform in directions $\beta := \frac{x}{r}$.

The assumptions concerning the smoothness of S and the continuity of $\zeta(s)$ can be considerably relaxed. This will be done later, in Chapter 2. The main results concerning existence and uniqueness of the scattering solutions and their properties. In particular, one is interested in the asymptotic behavior of the scattered field $v(x, \alpha, k)$ at the large distances from the scatterer D:

$$
v(x, \alpha, k) = \frac{e^{ikr}}{r} A(\beta, \alpha, k) + O\left(\frac{1}{r^2}\right), \quad |x| = r \to \infty, \quad \frac{x}{r} = \beta. \quad (1.7)
$$

Here $\beta \in S^2$ is the unit vector, the direction of the scattered wave, and $\alpha \in S^2$ is the unit vector, the direction of the incident plane wave $u_0(x, \alpha, k)$. If condition (1.6) holds, then the scattering solution exists and is unique, and the scattered field satisfies the asymptotic behavior (1.7). The function $A(\beta, \alpha, k)$ is called the scattering amplitude. It is the basic quantity describing the physical process of wave scattering. We study this function in detail in Chapter 2. The basic results of Chapter 2 are taken essentially from [9]; see also [5].

In Chapter 3 the inverse scattering problem is studied this problem consists of finding S and $\zeta(s)$ from the knowledge of $A(\beta, \alpha, k)$. If some of the variables is fixed, we denote it by subindex 0. For example, if $A(\beta, \alpha, k)$ is known for all $\beta, \alpha \in S^2$ and a fixed $k = k_0 > 0$, then the scattering data $A(\beta, \alpha, k) := A(\beta, \alpha)$ are called the fixed-energy (or fixed-frequency) scattering data. These data are four-dimensional since unit vectors α and β are described by two parameters (variables) each.

We also consider the data $A(\beta, \alpha_0, k) := A(\beta, k)$, known for all $\beta \in S^2$ and all $k > 0$. These data are three-dimensional.

A surface S in $\mathbb{R}^3$ is described by a function of two variables. In this sense S is a two-dimensional object.

The scattering data for an obstacle scattering is called *non-over-determined* if it is a function of two variables, for example, such is the data $A(\beta) := A(\beta, \alpha_0, k_0)$ known for all $\beta \in S^2$.

Let S_0^2 (or S_1^2) be an arbitrary fixed open subset of S^2.

The basic result of Chapter 3 is the following theorem.

Theorem 1.1 *The non-over-determined data $A(\beta)$, known for all $\beta \in S_0^2$, determine S and $\zeta(s)$ uniquely.*

The known results of this type were published only in the earlier works of the author [14], [9], [10], [26].

Theorem 1.1 is of principal importance because the non-over-determined scattering data are the minimal data which allow one to determine uniquely the unknown object, the surface S, and the boundary condition on S.

We prove in Chapter 3 the earlier result of the author; see [5].

Theorem 1.2 *The data $A(\beta, \alpha)$ known for all $\beta \in S_0^2, \alpha \in S_1^2$, determine S and $\zeta(s)$ uniquely.*

The fixed-energy scattering data $A(\beta, \alpha)$, used in Theorem 1.2, are of practical interest, but they are over-determined: they are four-dimensional while the unknown surface S is two-dimensional.

We also prove in Chapter 3 a generalization of the M. Schiffer's result which was not published by him. M. Schiffer proved that the data $A(\beta, k)$ known for all $\beta \in S^2$, all $k > 0$ determine S uniquely if the boundary condition on S is assumed to be the Dirichlet condition:

$$u = 0 \quad \text{on } S. \tag{1.8}$$

This corresponds to the case $\zeta(s) = \infty$ on S. The Schiffer's idea of the proof was presented in [5] with some generalizations: the data $A(\beta, k)$ were assumed known for $\beta \in S_0^2$ and k belonging to an arbitrary fixed open subset (a, b) of $[0, \infty)$. The boundary condition Γu was assumed of the type

$$\Gamma u = 0 \quad \text{on } S, \tag{1.9}$$

where

$$\Gamma u = u, \tag{1.10}$$

or

$$\Gamma u = u_N, \tag{1.11}$$

or

$$\Gamma u = u_N + \zeta(s)u. \tag{1.12}$$

There were two novel points in [5].

First, the usage of the analytic properties of $A(\beta, \alpha, k)$ allows to reduce the scattering data to their values on arbitrary fixed open subsets $\beta \in S_0^2, k \in (a, b)$.

Second, not only S but S and the boundary condition on S were uniquely determined by the scattering data.

Theorem 1.3   *The data $A(\beta, k), \beta \in S_0^2, k \in (a, b)$, determine S and the boundary condition on S uniquely.*

Theorem 1.1 allows one to develop a numerical method for solving inverse obstacle scattering problem with non-over-determined data. This was done for the first time by the author. A brief presentation of this result is given in Chapter 3, Section 3.5.

Finally, we formulate a new result, proved recently by the author. A proof is given in Lemma 3.6 in Chapter 3. This result is formulated in Theorem 1.4.

Theorem 1.4   *The scattering solution $u(x, \alpha, k)$ cannot have a closed surface of zeros different from S, the surface of the obstacle, if $u(s, \alpha, k) = 0$ for all $s \in S$.*

CHAPTER 2

The Direct Scattering Problem

2.1   STATEMENT OF THE PROBLEM

Let $D \subset \mathbb{R}^3$ be a bounded domain with a C^2-smooth connected boundary S, N be an outer unit normal to S, $D' = \mathbb{R}^3 \setminus D$ be the unbounded exterior region, $\zeta(s)$ be a continuous function on S, $\mathrm{Im}\zeta(s) \geq 0$, $N = N(s)$ be the unit normal on S, pointing into D', $k > 0$ is a constant, and S^2 is the unit sphere in $\mathbb{R}^3$, $\alpha, \beta \in S^2$.

The direct scattering problem by the obstacle D we call the obstacle scattering problem. This problem consists of finding the solution $u(x, \alpha, k)$ to the equation

$$\left(\nabla^2 + k^2\right)u \;=\; 0 \quad \text{in} \quad D', \tag{2.1}$$

$$\Gamma u \;=\; 0 \quad \text{on } S, \tag{2.2}$$

$$u \;=\; u_0 + v, \quad u_0 = e^{ik\alpha \cdot x}, \tag{2.3}$$

$$v_r - ikv \;=\; o\left(\frac{1}{r}\right), \quad r := |x| \to \infty, \tag{2.4}$$

and $o\left(\frac{1}{r}\right)$ is uniform in directions β of x, $\beta := \frac{x}{r}$. The boundary condition is assumed to be either the Dirichlet:

$$\Gamma u = u|_S = 0, \tag{2.5}$$

or Neumann:

$$\Gamma u = u_N|_S = 0, \tag{2.6}$$

or the impedance:

$$\Gamma u = \left(u_N + \zeta(s)u\right)|_S = 0, \quad \mathrm{Im}\zeta(s) \geq 0. \tag{2.7}$$

The impedance $\zeta(s)$ is a continuous function. The solution $u(x, \alpha, k)$ is called the scattering solution. The plane wave $u_0 = e^{ik\alpha \cdot x}$ is the incident field, α is the direction of the propagation of the incident field, $\alpha \cdot x$ is the dot product of two vectors, and $[\alpha, x]$ is the vector product of two vectors.

First, we prove that the solution of the direct scattering problem (2.1)–(2.4) is unique in the space $C(D')$ functions with the norm $||u|| = \max\limits_{x \in D'} |u(x)|$.

Second, we prove that this solution does exist in $C(D')$.

Third, we study properties of the scattering amplitude $A(\beta, \alpha, k)$. This amplitude is defined by the asymptotics of the scattered field:

$$v(x, \alpha, k) = \frac{e^{ikr}}{r} A(\beta, \alpha, k) + O\left(\frac{1}{r^2}\right), \quad r \to \infty, \quad \frac{x}{r} = \beta. \tag{2.8}$$

For bounded obstacle D, $o\left(\frac{1}{r}\right)$ in (2.4) can be replaced by $O\left(\frac{1}{r^2}\right)$.

Properties of the scattering amplitude were studied extensively in the literature, for example, in [9], [5], and will be presented in this chapter.

2.2   UNIQUENESS OF THE SCATTERING SOLUTION

Let us prove the following uniqueness result.

Theorem 2.1   *Problem* (2.1)–(2.4) *has at most one solution in* $C(D')$.

Proof of this theorem requires the lemma, known as Rellich's lemma.
Let $B_R = \{x : |x| \leq R\}$, $B'_R = \mathbb{R}^3 \setminus B_R$, $r := |x|$.

Lemma 2.2   *Any solution to Equation* (2.1) *in* B'_R *such that* $u(x) = o\left(\frac{1}{r}\right)$ *is equal to zero in* B'_R. *The same is true if* $\int_{S^2} |u(r, x^0)|^2 dx^0 = o\left(\frac{1}{r^2}\right), r = |x| \to \infty$.

Proof of Lemma 2.2. (see [5] or [9], p. 30).

By separation of variables in the spherical coordinates one obtains:

$$u(x) = \sum_{l=0}^{\infty} c_l h_l^{(1)}(r) Y_l(x^0),$$

where c_l are constants, $r = |x|, x^0 = \frac{x}{r}, h_l^{(1)}(r)$ are the spherical Hankel functions satisfying the radiation condition (2.4), and $Y_l(x^0)$ are the normalized spherical harmonics; see, for example, [5] or [9]. If $u(x) = o\left(\frac{1}{|x|}\right)$, then one gets a contradiction unless $c_l = 0$ for all $l \geq 0$. This contradiction is a consequence of the formula

$$\frac{1}{r^2} \sum_{l=0}^{\infty} |c_l|^2 = o\left(\frac{1}{r^2}\right) \text{ as } r \to \infty,$$

which implies

$$\sum_{l=0}^{\infty} |c_l|^2 = 0.$$

Consequently, $c_l = 0$ for all $l \geq 0$ and $u = 0$.
Lemma 2.2 is proved. $\square$

We will use also the unique continuation theorem for solutions u of Equation (2.1).

Lemma 2.3   *If $u = 0$ on a set of positive Lebesque measure in D', then $u = 0$ in D'.*

Lemma 2.3 is a consequence of the analyticity of the solutions to (2.1).

Proof of Theorem 2.1. This theorem is equivalent to the statement that any solution $v \in C(D')$ to the homogeneous problem (2.1)–(2.4), that is, this problem with $u_0 = 0$, is equal to zero identically. To prove this, consider the identity

$$0 = \int_{B_r \setminus D} \bar{v} \left(\nabla^2 + k^2 \right) v dx = \int_{S_R} \bar{v} v_r ds - \int_S \bar{v} v_N ds + \int_{B_R \setminus D} k^2 |v|^2 dx, \qquad (2.9)$$

where S_R is the boundary of B_R, $B_R \supset D$.

Using the radiation condition (2.4) one gets

$$\int_{S_R} \bar{v} v_r ds = ik \int_{S_R} |v|^2 ds + o(1), \quad R \to \infty. \qquad (2.10)$$

Using the boundary condition (2.7) one gets

$$-\int_S \bar{v} v_N ds = \int_S \zeta(s) |v|^2 ds. \qquad (2.11)$$

Taking the imaginary part of (2.9) and using (2.10)–(2.11) yields

$$0 = \int_{S_R} k |v|^2 ds + \int_S \operatorname{Im}\zeta(s) |v|^2 ds + o(1), \quad R \to \infty. \qquad (2.12)$$

Since $\operatorname{Im}\zeta(s) \geq 0$ and $k > 0$, Equation (2.12) implies

$$\lim_{R \to \infty} \int_{S_R} |v|^2 ds = 0 \qquad (2.13)$$

(and $v(s) = 0$ at the points s at which $\operatorname{Im}\zeta(s) > 0$.)

By Lemma 2.2, it follows from (2.13) that $v = 0$ in B'_R.

By Lemma 2.3, $v = 0$ in D'.

Theorem 2.1 is proved.   □

The above proof is valid for the Dirichlet boundary condition as well.

2.3   EXISTENCE OF THE SCATTERING SOLUTION

Our proof of the existence of the scattering solution is based on the reduction of the existence problem to solving a Fredholm integral equation. The homogeneous version of this equation has only the trivial solution by Theorem 2.1. Therefore, by the Fredholm alternative, this equation has a solution and the solution is unique.

Let us derive the above-mentioned Fredholm equation. Look for the scattered field of the form

$$v(x) = \int_S g(x,s)\sigma(s)ds := T\sigma, \quad g(x,y) = \frac{e^{ik|x-y|}}{4\pi|x-y|}, \tag{2.14}$$

where $\sigma(s)$ is an unknown continuous function and the operator T is a linear compact operator in $L^2(S)$ and in $C(S)$. It is known (see, for example, [5] or [9], p. 18) that

$$\frac{\partial v}{\partial N_{\pm}} = \frac{A\sigma \pm \sigma}{2}, \tag{2.15}$$

where

$$A\sigma = 2\int_S \frac{\partial g(s,t)}{\partial N_s}\sigma(t)dt, \tag{2.16}$$

and $\frac{\partial}{\partial N_+}\left(\frac{\partial}{\partial N_-}\right)$ denotes the limiting value of the normal derivative from inside (respectively, outside) of D.

It is also well known that the operator A is a compact operator in the spaces $L^2(S)$ and $C(S)$ if S is C^2-smooth (and under less restrictive assumptions on S).

The boundary condition (2.7) together with formulas (2.14)–(2.16) yields:

$$u_{0N} + \frac{A\sigma - \sigma}{2} + \zeta(s)(u_0 + T\sigma) = 0. \tag{2.17}$$

Let us rewrite Equation (2.17) as

$$\sigma(s) = A\sigma + 2\zeta(s)T\sigma + 2\zeta(s)u_0(s) + 2u_{0N}(s). \tag{2.18}$$

The properties of the operators T and A were discussed in detail in the literature, for example, in [20].

Since A and T are linear compact operators in $L^2(S)$ or $C(S)$, and the free term in (2.18) belongs to both of these spaces if S is C^2-smooth, one concludes that Equation (2.18) is of Fredholm type. Its homogeneous version

$$\sigma = A\sigma + 2\zeta(s)T\sigma \tag{2.19}$$

is equivalent to the relation

$$\frac{\partial v}{\partial N_-} + \zeta(s)v = 0 \quad \text{on } S. \tag{2.20}$$

The function v, defined by (2.14), satisfies equation

$$(\nabla^2 + k^2)v = 0 \quad \text{in } D'. \tag{2.21}$$

It also satisfies the radiation condition (2.4) and the homogeneous boundary condition (2.20) with $\mathrm{Im}\zeta(s) \geq 0$. By Theorem 2.1 it follows that $v = 0$ in D'.

Let us prove that if $v = 0$ in D' then $\sigma = 0$.

Indeed, by continuity of the single layer potential v up to the boundary S it follows that if $v(x) = 0$ in D', then $v|_S = 0$ and $v_{N_-}|_S = 0$.

It follows from (2.15) that

$$\sigma = v_{N_+} - v_{N_-}. \tag{2.22}$$

Therefore $\sigma = 0$ if $v = 0$ in D.

A single layer potential which vanishes on S may be not zero in D if k^2 is a Dirichlet eigenvalue of the operator $-\Delta$ in D. Therefore, if k^2 is not a Dirichlet eigenvalue of the operator $-\Delta$ in D, then $\sigma = 0$ and the proof is complete.

If k^2 is a Dirichlet eigenvalue of the operator $-\Delta$ in D then our proof should be changed. In [5] and [9] several methods are given for deriving a Fredholm integral equation for a scattering solution so that this integral equation is *unconditionally solvable*.

Let us describe one of these methods. Choose a small ball $B_\epsilon \subset D$ such that k^2 is not a Dirichlet eigenvalue of the operator $-\Delta$ in the domain $D_\epsilon := D \setminus B_\epsilon$. This is possible to do by choosing suitable $\epsilon > 0$.

Lemma 2.4   *For any $k > 0$, there exists an $\epsilon > 0$ such that the problem*

$$\begin{aligned}
\left(\nabla^2 + k^2\right) v &= 0 & &\text{in } D \setminus B_\epsilon, & (2.23)\\
v &= 0 & &\text{on } S_\epsilon = \partial B_\epsilon, & (2.24)\\
v &= 0 & &\text{on } S = \partial D & (2.25)
\end{aligned}$$

has only the trivial solution $v = 0$. There are infinitely many such $\epsilon > 0$.

Proof. The conclusion of Lemma 2.4 is a consequence of Lemma 2.5. $\qquad\square$

Lemma 2.5   *Let $k_1^2(\epsilon) \leq k_2^2(\epsilon) \leq \cdots \leq k_n^2(\epsilon) \leq \ldots$ be the eigenvalues of the problem (2.23)–(2.25) counted according to their multiplicities. If $\epsilon_1 < \epsilon_2$, then $k_n^2(\epsilon_1) < k_n^2(\epsilon_2)$.*

Proof. If $a \leq b$ then the variational definition of the eigenvalues implies $k_n^2(a) \leq k_n^2(b)$. Suppose that $a < b$ and $k_n^2(a) = k_n^2(b)$. If $a < \epsilon < b$, then $k_n^2(a) \leq k_n^2(\epsilon) \leq k_n^2(b)$, so $k_n^2(a) = k_n^2(\epsilon) := k^2$ for any $\epsilon \in [a, b]$. Let us choose $a < \epsilon_1 < \epsilon_2 < \cdots < \epsilon_m < b$ with as large m as we wish. Then $k_n^2(\epsilon_j) = k^2, 1 \leq j \leq m$. To each j corresponds the normalized eigenfunction $u_j(x)$ of the Dirichlet Laplacian in $D \setminus B_{\epsilon_j}$ and k^2 is the corresponding eigenvalue. Define $u_j(x) = 0$ in $B_{\epsilon_j} \setminus B_{\epsilon_1}$. Then $u_j \in H^1(D \setminus B_{\epsilon_1})$, and the system $\{u_j(x)\}_{j=1}^m$ is linearly independent in $L^2(D \setminus B_{\epsilon_1})$ and in $C(D \setminus B_{\epsilon_1})$. The linear independence is easily checked:

if $\sum_{j=1}^{m} c_j u_j(x) = 0$, then take $x \in B_{\epsilon_2} \setminus B_{\epsilon_1}$ and get $c_1 = 0$. Continue in this fashion and get all $c_j = 0, 1 \leq j \leq m$. Let $u(x) = \sum_{j=1}^{m} c_j u_j(x)$. Choose the coefficients c_j so that

$$(u, u_j) := \int_{D \setminus B_{\epsilon_1}} u(x) \overline{u_j(x)} dx = 0, \quad 1 \leq j \leq m-1, \tag{2.26}$$

$$||u_j||^2 = (u_j, u_j) = 1. \tag{2.27}$$

By the variational principle for eigenvalues one has

$$\inf_{w \perp L_{m-1}} \frac{(Aw, w)}{(w, w)} \geq \lambda_m. \tag{2.28}$$

Here A is a non-negative self-adjoint operator with a discrete spectrum, L_{m-1} is an m-dimensional linear subspace in the domain of the quadratic form of A, $D(A^{1/2})$. If one takes by L_m a linear span of $u_j, 1 \leq j \leq m-1$, and by A the Dirichlet operator $-\Delta$ in $D \setminus B_{\epsilon_1}$, then for $u = \sum_{j=1}^{m} c_j u_j$ one has:

$$\frac{(\nabla u, \nabla u)}{(u, u)} \geq k_m^2. \tag{2.29}$$

On the other hand, for $||u|| = 1$, one has

$$\frac{(\nabla u, \nabla u)}{(u, u)} = \sum_{j,i=1}^{m} c_j \bar{c}_i (\nabla u_j, \nabla u_i) = \sum_{j,i=1}^{m} c_j \bar{c}_i k^2 \delta_{ji} = k^2. \tag{2.30}$$

For sufficiently large m one has $k_m^2 > k^2$. Therefore, (2.29) and (2.30) lead to a contradiction. This contradiction proves Lemma 2.5. □

See also [5, pp. 29–30].

It is now possible to derive an unconditionally solvable integral equation for finding the scattering solution in the case when k^2 is a Dirichlet eigenvalue of the Laplacian in D. Namely, let $g(x, y)$ be the Green's function of the Dirichlet Laplacian in $\mathbb{R}^3 \setminus B_\epsilon$, where $\epsilon > 0$ is such that k^2 is not a Dirichlet eigenvalue of the Laplacian in $D \setminus B_\epsilon$. Such $\epsilon > 0$ does exist by Lemma 2.4. Let

$$v_\epsilon = \int_S g_\epsilon(x, s) \sigma(s) ds := T_\epsilon \sigma. \tag{2.31}$$

Then formulas (2.15)–(2.18) are valid with A_ϵ replacing A and T_ϵ replacing T,

$$A_\epsilon \sigma = 2 \int_S \frac{\partial g_\epsilon(s, t)}{\partial N_s} \sigma(t) dt. \tag{2.32}$$

The compactness properties of A_ϵ and T_ϵ as operators in $L^2(S)$ or in $C(S)$ are the same as these of A and T. Therefore, Equation (2.18) with A_ϵ in place of A and T_ϵ in place of T is a Fredholm integral equation.

Its homogeneous version, corresponding to $u_0 = 0$, has only the trivial solution. Indeed, as above, one proves that $v_\epsilon = 0$ in D', $v_\epsilon|_S = 0$ and $\frac{\partial v_\epsilon}{\partial N_-}|_S = 0$. Since v_ϵ solves Equation (2.21) in $D \setminus B_\epsilon$ and $v_\epsilon|_S = 0 = v_\epsilon|_{S_\epsilon}$, it follows that $v_\epsilon = 0$ in $D \setminus B_\epsilon$ because k^2 is not a Dirichlet eigenvalue of the Laplacian in $D \setminus B_\epsilon$. Therefore, $v_\epsilon = 0$ in $D \setminus B_\epsilon$ and in D'. Consequently, $\frac{\partial v_\epsilon}{\partial N_\pm} = 0$ on S.

By a formula similar to (2.22) one concludes that the null space of the operator $A_\epsilon - 2\zeta(s)T_\epsilon$ is trivial. Therefore, the scattering solution does exist and is unique. Moreover, it can be found in the form $u = u_0 + v$, where v is defined by formula (2.14) if $k^2 > 0$ is not a Dirichlet eigenvalue of the Laplacian in D, or in the form $u = u_0 + v_\epsilon$, where v_ϵ is defined by formula (2.31) if $k^2 > 0$ is a Dirichlet eigenvalue of the Laplacian in D.

Let us formulate our basic result.

Theorem 2.6   *The solution to problem* (2.1)–(2.4) *does exist and is unique.*

Actually, we have proved more. Namely, we have proved that the scattering solution $u(x, \alpha, k)$ can be found in the form (2.3) with v defined either in (2.14) or in (2.31) depending on whether k^2 is not a Dirichlet eigenvalue of the Laplacian in D or is such an eigenvalue.

Let us study the behavior of the scattered field v as $|x| = r \to \infty$, $\frac{x}{r} = \beta$.

Let $G(x, y, k)$ be the Green's function which is the unique solution of the problem

$$\left(\nabla^2 + k^2\right) G(x, y, k) = -\delta(x - y) \quad \text{in } D', \tag{2.33}$$

$$\frac{\partial G(x, y, k)}{\partial N} - ikG(x, y, k) = o\left(\frac{1}{|x|}\right), \quad |x| \to \infty, \tag{2.34}$$

$$\frac{\partial G}{\partial N} + \zeta(s)G = 0 \quad \text{on } S, \quad \text{Im } \zeta(s) \ge 0. \tag{2.35}$$

By Green's formula one derives

$$G(x, y, k) = g(x, y, k) + \int_S G(x, s, k) \left(\frac{\partial}{\partial N_s} + \zeta(s)\right) g(s, y, k)ds. \tag{2.36}$$

Let us take $|y| \to \infty$, $\frac{y}{|y|} = -\beta$ in formula (2.36). Then one gets

$$G(x, y, k) = \frac{e^{ik|y|}}{4\pi|y|}u(x, \alpha, k) + o\left(\frac{1}{|y|}\right), \quad |y| \to \infty, \quad \frac{y}{|y|} = -\alpha, \tag{2.37}$$

where

$$u(x, \alpha, k) = e^{ik\alpha \cdot x} + \int_S G(x, s, k) \left(\frac{\partial}{\partial N_s} + \zeta(s)\right) e^{ik\alpha \cdot s}ds. \tag{2.38}$$

Note that $u(x, \alpha, k)$ solves Equation (2.1), is of the form (2.3) and satisfies the radiation condition (2.4). Let us check that the boundary condition (2.7) is satisfied. If this is done, then $u(x, \alpha, k)$ is the scattering solution which is unique by Theorem 2.6.

Let u be defined by (2.38). Apply the Green's formula to the equations

$$\begin{aligned}
(\nabla^2 + k^2)\, u &= 0 \quad \text{in } D', & (2.39)\\
(\nabla^2 + k^2)\, G &= -\delta(x-y) \quad \text{in } D', & (2.40)
\end{aligned}$$

and get

$$u = \int_{S_R} [G(x,s)u_N - G_N(x,s)u]\, ds - \int_S [G(x,s)u_N - G_N(x,s)u]\, ds. \tag{2.41}$$

One has $-G_N = \zeta(s)G$ on S, and $u_N + \zeta(s)u = 0$ on S. So,

$$\int_S G(x,s)\,(u_N + \zeta(s)u)\, ds = 0. \tag{2.42}$$

Furthermore,

$$\int_{S_R} [Gu_N - G_N u]\, ds = \int_{S_R} [Gu_{0N} - G_N u_0]\, ds + \int_{S_R} [Gv_N - G_N v]\, ds. \tag{2.43}$$

By the radiation condition one has:

$$\lim_{R\to\infty} \int_{S_R} [Gv_N - G_N v]\, ds = 0. \tag{2.44}$$

Consequently,

$$\begin{aligned}
\lim_{R\to\infty} \int_{S_R} [Gu_{0N} - G_N u_0]\, ds &= \lim_{R\to\infty} \int_{B_R\setminus D} \left[G\left(\nabla^2 + k^2\right)u_0 - \left(\nabla^2 + k^2\right)Gu_0 \right] ds \\
&+ \int_S [Gu_{0N} - G_N u_0]\, ds = u_0(x) + \int_S G(x,s,k)\left(\frac{\partial}{\partial N_s} + \zeta(s)\right)u_0 ds = u(x,\alpha,k),
\end{aligned} \tag{2.45}$$

where we used formula (2.38).

Therefore, we have checked that the solution to Equation (2.39) in D', satisfying the boundary condition (2.7) and conditions (2.3)–(2.4) is the solution defined by formula (2.38).

Let us formulate the result we have proved; see also [8].

Lemma 2.7   *[Ramm's lemma] Formula* (2.37) *holds, where* $u(x,\alpha,k)$ *is the scattering solution.*

2.4   PROPERTIES OF THE SCATTERING AMPLITUDE

Define the scattering amplitude $A(\beta,\alpha,k)$ by the relation:

$$v(x,\alpha,k) = \frac{e^{ik|x|}}{|x|} A(\beta,\alpha,k) + o\left(\frac{1}{|x|}\right), \quad |x| \to \infty, \quad \frac{x}{|x|} = \beta. \tag{2.46}$$

From this definition and from formulas (2.37)–(2.38) one derives

$$A(\beta, \alpha, k) = \frac{1}{4\pi} \int_S u(s, -\beta, k) \left(\frac{\partial}{\partial N_s} + \zeta(s) \right) e^{ik\alpha \cdot s} ds, \tag{2.47}$$

where $u(x, -\beta, k)$ is the scattering solution and

$$v(x, \alpha, k) = \int_S G(x, s, k) \left(\frac{\partial}{\partial N_s} + \zeta(s) \right) e^{ik\alpha \cdot s} ds, \quad x \in D'. \tag{2.48}$$

It follows from formula (2.47) that $A(\beta, \alpha, k)$ is an analytic function of $\alpha \in \mathbb{C}^3$ on the analytic algebraic variety defined by the equation:

$$\alpha \cdot \alpha = \sum_{j=1}^{3} \alpha_j^2 = 1. \tag{2.49}$$

In the usual way one proves the symmetry of G:

$$G(x, y, k) = G(y, x, k), \quad x, y \in D'. \tag{2.50}$$

We will use the following lemma, proved in [9, p. 62].

Let

$$\gamma(r) := \frac{e^{ikr}}{r}. \tag{2.51}$$

Lemma 2.8 *If $f \in C^1(S^2)$, then*

$$\int_{S^2} e^{ik\alpha \cdot \omega r} f(\omega) d\omega = \frac{2\pi i}{k} [\bar{\gamma}(r) f(-\alpha) - \gamma(r) f(\alpha)] + o\left(\frac{1}{r}\right), \quad r \to \infty. \tag{2.52}$$

This formula can be written as

$$e^{ikr\alpha \cdot \omega} \sim \frac{2\pi i}{k} [\bar{\gamma}(r)\delta(\omega + \alpha) - \gamma(r)\delta(\omega - \alpha)], \quad r \to \infty, \tag{2.53}$$

where $\sim$ stands for the asymptotic equality and $\delta(\omega)$ is the delta function on S^2.

Proof. Let $\omega = (\theta, \phi), \alpha \cdot \omega = \cos \theta$, so α is directed along z-axis. Let $\xi = \cos \theta, \theta = \arccos \xi$. One has

$$\int_{S^2} e^{ikr\alpha \cdot \omega} f(\omega) d\omega = \int_0^{2\pi} d\phi \int_{-1}^{1} d\xi e^{ikr\xi} f(\arccos \xi, \phi) =$$

$$= \int_0^{2\pi} d\phi \left[2\pi \frac{e^{ikr}}{ikr} f(\alpha) - \frac{e^{-ikr}}{ikr} f(-\alpha) + o\left(\frac{1}{r}\right) \right] = \tag{2.54}$$

$$= \frac{2\pi}{ik} [\gamma(r) f(\alpha) - \bar{\gamma}(r) f(-\alpha)] + o\left(\frac{1}{r}\right), \quad r \to \infty.$$

This formula coincides with formula (2.52). It can be written as formula (2.53). Lemma 2.8 is proved.   □

Lemma 2.9   *One has*

$$A(\beta, \alpha, k) = A(-\alpha, -\beta, k). \tag{2.55}$$

Proof. Let $u(\alpha) := u(x, \alpha, k)$ be the scattering solution, $u_0(x, \alpha, k) = e^{ikr\omega\cdot\alpha}$, $r = |x|$, $\omega = \frac{x}{r}$,

$$u(\alpha) := u(x, \alpha, k) = u_0(x, \alpha, k) + v(\alpha)$$

$$= u_0 + \gamma A(\omega, \alpha, k) + o\left(\frac{1}{r}\right), \ |x| = r \to \infty, \ \frac{x}{r} = \omega, \tag{2.56}$$

$$u(\beta) := u(x, \beta, k) = u_0(x, \beta, k) + v(\beta)$$

$$= u_0 + \gamma A(\omega, \beta, k) + o\left(\frac{1}{r}\right), \ |x| = r \to \infty, \ \frac{x}{r} = \omega, \tag{2.57}$$

$$(\nabla^2 + k^2)\, u(\alpha) = 0, \quad (\nabla^2 + k^2)\, u(\beta) = 0. \tag{2.58}$$

By Green's formula one gets:

$$0 = \int_{B_R \setminus D} \left[u(\alpha)\,(\nabla^2 + k^2)\, u(\beta) - u(\beta)\,(\nabla^2 + k^2)\, u(\alpha)\right] dy$$

$$= \int_{S_R} [u(\alpha)u_N(\beta) - u(\beta)u_N(\alpha)]\, ds - \int_S [u(\alpha)u_N(\beta) - u(\beta)u_N(\alpha)]\, ds. \tag{2.59}$$

By the boundary conditions

$$u_N(\alpha) + \zeta(s)u = 0, \quad u_N(\beta) + \zeta(s)u = 0 \quad \text{on } S, \tag{2.60}$$

the last integral in (2.59) equals to zero. Thus, (2.59) can be written as

$$0 = \int_{S_R} [u_0(\alpha)u_{0N}(\beta) - u_0(\beta)u_{0N}(\alpha)]\, ds + \int_{S_R} [v(\alpha)v_N(\beta) - v(\beta)v_N(\alpha)]\, ds$$

$$+ \int_{S_R} [u_0(\alpha)v_N(\beta) - v(\beta)u_{0N}(\alpha)]\, ds + \int_{S_R} [v(\alpha)u_{0N}(\beta) - u_0(\beta)v_N(\alpha)]\, ds. \tag{2.61}$$

The first integral in (2.61) equals to zero since $u_0(\alpha)$ and $u_0(\beta)$ solve Equation (2.58) in the whole space. The second integral in (2.61) equals to zero as $R \to \infty$ since $v(\alpha)$ and $v(\beta)$ satisfy the radiation condition

$$v_r - ikv = o\left(\frac{1}{r}\right), \quad v = \gamma(r)A(\omega, \alpha, k) + o\left(\frac{1}{r}\right), \ r \to \infty \tag{2.62}$$

and $v_N = v_r$ on S_R. We drop the k-dependence in the rest of the proof to save space.

Therefore, (2.61) can be rewritten as

$$0 = \lim_{R \to \infty} \left\{ \int_{S_R} [u_0(\alpha)v_r(\beta) - v(\beta)u_{0r}(\alpha)]ds + \int_{S_R} [v(\alpha)u_{0r}(\beta) - v_r(\alpha)u_0(\beta)ds \right\}. \quad (2.63)$$

One has

$$v_r(\beta) = ikv(\beta) + o\left(\frac{1}{R}\right), \quad v_r(\alpha) = ikv(\alpha) + o\left(\frac{1}{R}\right), \quad R \to \infty, \quad (2.64)$$

$$v(\beta) = \gamma(r)A(\omega, \beta) + o\left(\frac{1}{R}\right), \quad R \to \infty. \quad (2.65)$$

By formula (2.53) one gets

$$\int_{S_R} [u_0(\alpha)v_r(\beta) - v(\beta)u_{0r}(\alpha)]ds$$

$$\sim \int_{S_R} \left[e^{ikR\omega \cdot \alpha} ikA(\omega, \beta)\gamma(R) - \gamma(R)A(\omega, \beta)ike^{ikR\omega \cdot \alpha}\omega \cdot \alpha \right] ds$$

$$\sim ik\gamma(R)R^2 \left\{ \left[\frac{2\pi i}{k}\bar{\gamma}(R)A(-\alpha, \beta) - \frac{2\pi i}{k}\gamma(R)A(\alpha, \beta) \right] - \quad (2.66) \right.$$

$$\left. - \left[\frac{2\pi i}{R}\bar{\gamma}(R)A(-\alpha, \beta)(-1) - \frac{2\pi i}{k}\gamma(R)A(\alpha, \beta) \right] \right\}$$

$$= -2\pi A(-\alpha, \beta) + 2\pi e^{2ikR}A(\alpha, \beta) - 2\pi A(-\alpha, \beta) - 2\pi e^{2ikR}A(\alpha, \beta)$$

$$= -4\pi A(-\alpha, \beta).$$

Similarly,

$$\int_{S_R} [v(\alpha)u_{0r}(\beta) - v_r(\alpha)u_0(\beta)] ds$$

$$\sim \int_{S_R} \left[\gamma(R)A(\omega, \alpha)ike^{ikR\omega \cdot \beta}\omega \cdot \beta - ik\gamma(R)A(\omega, \alpha)e^{ikR\omega \cdot \beta} \right] ds$$

$$\sim ik\gamma(R)R^2 \quad (2.67)$$

$$\left\{ \frac{2\pi i}{k} [-\bar{\gamma}(R)A(\beta, \alpha) - \gamma(R)A(\beta, \alpha)] - \frac{2\pi i}{k} [\bar{\gamma}(R)A(-\beta, \alpha) - \gamma(R)A(\beta, \alpha)] \right\}$$

$$= 4\pi A(-\beta, \alpha).$$

From (2.63), (2.66), and (2.67) one obtains

$$0 = -A(-\alpha, \beta, k) + A(-\beta, \alpha, k). \quad (2.68)$$

This formula is equivalent to (2.55).

Lemma 2.9 is proved.   □

Lemma 2.10   *The scattering amplitude $A(\beta, \alpha, k)$ is an analytic function of β and α on the algebraic variety $Z_1 = \{\alpha : \alpha \in \mathbb{C}^3, \alpha \cdot \alpha = 1\}$, where*

$$\alpha \cdot \alpha = \sum_{j=1}^{3} \alpha_j^2, \quad \alpha_j \in \mathbb{C}. \tag{2.69}$$

Proof. From formula (2.47) it follows that $A(\beta, \alpha, k)$ is an analytic function on the variety $\alpha \in Z_1$ because $e^{ik\alpha \cdot s}$ is such a function. By formula (2.55) it follows that $A(\beta, \alpha, k)$ is also an analytic function of $\beta \in Z_1$.

Lemma 2.10 is proved.   □

Lemma 2.11   *The scattering amplitude $A(\beta, \alpha, k)$ is an analytic function of k in $\mathrm{Im}k \geq 0$ and a meromorphic function of k in $\mathrm{Im}k < 0$.*

Proof. By Lemma 2.7 the scattering solution is locally an analytic function of k in the domains where $G(x, y, k)$ is an analytic function of k. The scattering amplitude is an analytic function of k in the domains in which the scattering solution is such a function. For $\mathrm{Im}k \geq 0$ the scattering solution satisfies a Fredholm integral equation, the homogeneous version of which has only the trivial solution and the kernel of which is an analytic function of k in a neighborhood of any k with $\mathrm{Im}k \geq 0$. This implies that $A(\beta, \alpha, k)$ is an analytic function of k in the half-plane $\mathrm{Im}k \geq 0$.

The full conclusion of Lemma 2.11 follows from the theorem proven in [9, pp. 64–69]. Let $R(B)$ be the range of a linear operator B and $N(B)$ be the null space of B.

Theorem 2.12   *Let $B(k)$ be a bounded mapping of a Banach space X into a Banach space Y. Assume that $B(k)$ is an analytic function of k in a connected domain Δ of the complex plane $\mathbb{C}$, and $B(k) \in Fred(X, Y)$, that is, $R(B(k))$ is closed and $r := dimN(B(k)) = dimN((B(k))^*) < \infty$. Then either $B^{-1}(k)$ does not exist at any point $k \in \Delta$, or $B^{-1}(k)$ exists at all points $k \in \Delta$ except for a discrete set of points k_j. In a neighborhood of any $k \notin k_j$ the operator $B^{-1}(k)$ is an analytic function of k. The points k_j are poles of $B^{-1}(k)$. The corresponding Laurent coefficients $B_{pj}, -m_j \leq p \leq -1$ are finite-rank operators,*

$$B^{-1}(k) = \sum_{p=-m_j}^{\infty} B_{pj}(k - k_j)^p. \tag{2.70}$$

Lemma 2.11 is proved.   □

CHAPTER 3

Inverse Obstacle Scattering

3.1 STATEMENT OF THE PROBLEM

Let us make the same assumptions about obstacle D as in Section 2.1, when the direct obstacle scattering problem was discussed.

The inverse obstacle scattering problem consists of finding the surface S of the obstacle and the boundary condition on S from the scattering data $A(\beta, \alpha, k)$.

Since the unit vectors α and β are described by two variables (coordinates) each, the scattering data $A(\beta, \alpha, k)$ is a function of five variables. If some of the variables is fixed we denote this variable by index 0. For example, if $k = k_0$ then the values $A(\beta, \alpha, k) := A(\beta, \alpha)$ are fixed-energy scattering data which are four-dimensional. If the incident direction is fixed, that is, $\alpha = \alpha_0$, then the values $A(\beta, \alpha_0, k) := A(\beta, k)$ are three-dimensional scattering data. The surface S of the obstacle D can be described by a function of two variables. We say that the unknown S is a two-dimensional object and the data $A(\beta, \alpha)$ are over-determined, that is, they are functions of more than two variables.

The scattering data $A(\beta, \alpha_0, k_0) := A(\beta)$ are called non-over-determined since $A(\beta)$ is a function of two variables, two-dimensional scattering data.

It is easy to prove that the five-dimensional scattering data $A(\beta, \alpha, k)$ known for all $\beta, \alpha \in S^2$ and all $k > 0$ determine the surface S uniquely. If S is known then the boundary condition on S is uniquely determined as we prove later.

The first uniqueness result on finding S from the data $A(\beta, k)$ was obtained by M. Schiffer in 1964, but M. Schiffer did not publish this result. He assumed that the boundary condition on S was the Dirichlet condition. M. Schiffer's proof was published, for example, in [5], [9]. In [5] the boundary condition on S was not assumed known *a priori*, but was determined uniquely by the scattering data. In [5] the uniqueness theorem for the inverse obstacle scattering with fixed-energy data $A(\beta, \alpha)$ was proved for the first time. Not only S was determined uniquely but also the boundary condition on S. It was assumed that the boundary condition on S is either the Dirichlet type, or Neumann, or the impedance boundary condition. For a long time there were no uniqueness results for inverse obstacle scattering problem with non-over-determined data. Such results were obtained by the author in [14], [9], [10], [21], [22], [26]. These results are presented in this chapter.

The statement of inverse obstacle scattering problem is simple:

Given the scattering data, can one find uniquely the surface S and the boundary condition on S?

Lemma 3.1   *If the scattering data $A(\beta, \alpha)$ are known for all $\beta \in S_1^2$ and all $\alpha \in S_2^2$, where S_j^2, $j = 1, 2$, are arbitrary fixed open subsets of S^2, then $A(\beta, \alpha)$ are uniquely determined for all $\beta \in S^2$ and all $\alpha \in S^2$. If the scattering data $A(\beta, k)$ are known for all $\beta \in S_1^2$ and all $k \in (a, b), 0 \leq a < b$, then $A(\beta, k)$ is uniquely determined for all $\beta \in S^2$ and all $k > 0$.*

Proof. The conclusions of Lemma 3.1 follow immediately from Lemma 2.11.   □

Corollary 3.2   *When we say that the scattering data $A(\beta, \alpha)$ are known for all α, β, then these data are known for all $\beta \in S_1^2$, and all $\alpha \in S_2^2$ and therefore these data are known for all $\beta \in S^2$ and all $\alpha \in S^2$.*

When we say that the data $A(\beta, k)$ are known for all β and all k, then these data are known for all $\beta \in S_1^2$ and all $k \in (a, b), 0 \leq a < b$, and therefore these data are known for all $\beta \in S^2$ and all $k > 0$.

3.2   UNIQUENESS OF THE SOLUTION TO OBSTACLE INVERSE SCATTERING PROBLEM WITH THE DATA $A(\beta)$

In this section the main result is formulated in the following theorem.

Theorem 3.3   *The data $A(\beta)$ determine uniquely the surface S and the boundary condition on S.*

Clearly, this theorem implies the uniqueness theorem for the inverse obstacle scattering problem with the over-determined scattering data $A(\beta, k)$ or $A(\beta, \alpha)$. However, we will discuss these cases separately later.

Proof. Suppose that there are two surfaces S_j, $j = 1, 2$, and two boundary conditions Γ_j, which produce the same scattering data $A(\beta, k)$. Let D_j, $j = 1, 2$, be the corresponding obstacles.

Denote $D_{12} := D_1 \cup D_2$, $D^{12} := D_1 \cap D_2$, let $S_{12} = \partial D_{12}$ be the boundary of D_{12}, and $S^{12} = \partial D^{12}$ be the boundary of D^{12}. Let $D_{12} \setminus D^{12} := D_3$ and $S_3 = \partial D_3$. Denote by D_4 a connected component of D_3 such that its complement $D_4' := \mathbb{R}^3 \setminus D_4$ is unbounded. By D' we denote the complement of D, $D' = \mathbb{R}^3 \setminus D$. By $u_j(x, \alpha, k) := u_j$, $j = 1, 2$, we denote the scattering solutions corresponding to D_j and by v_j the corresponding scattered fields. We consider three cases:

(a) D^{12} is empty set, that is, D_1 and D_2 do not intersect,

(b) $D^{12} = D_2$, that is, $D_2 \subset D_1$, and

(c) D^{12} is non-empty and does not coincide with one of the obstacles.

Case (a).

The scattering solution u_j is analytic in D'_j. Consider

$$w = u_1 - u_2 = v_1 - v_2. \tag{3.1}$$

By our assumption

$$v_1 - v_2 = A(\beta, k)\gamma(r) - A(\beta, k)\gamma(r) + o\left(\frac{1}{r}\right), \quad r := |x| \to \infty, \tag{3.2}$$

where

$$\gamma(r) := \frac{e^{ikr}}{r}. \tag{3.3}$$

Thus,

$$w = v_1 - v_2 = o\left(\frac{1}{r}\right), \quad r \to \infty. \tag{3.4}$$

The function w solves the equation

$$\left(\nabla^2 + k^2\right)w = 0, \quad |x| \geq R, \tag{3.5}$$

where $R > 0$ is sufficiently large so that

$$D_{12} \subset B_R, \quad B_R := \{x : |x| \leq R\}.$$

By Lemma 2.2 one concludes that conditions (3.4)–(3.5) imply

$$w = 0, \quad |x| \geq R. \tag{3.6}$$

By Lemma 2.3, that is, by the unique continuation property of the solutions to the Helmholtz Equation (3.5), it follows from (3.6) that

$$w = 0 \quad \text{in } D'_{12}. \tag{3.7}$$

Therefore,

$$v_1 = v_2 := v \quad \text{in } D'_{12}. \tag{3.8}$$

Since v_j is analytic in D'_j and D_{12} is empty, it follows from (3.8) that v is analytic in $\mathbb{R}^3$. Therefore,

$$\left(\nabla^2 + k^2\right)v \; = \; 0 \quad \text{in } \mathbb{R}^3, \quad k = \text{const} > 0, \tag{3.9}$$

$$v_r - ikv \; = \; o\left(\frac{1}{r}\right), \quad r \to \infty. \tag{3.10}$$

Lemma 3.4   *Any solution to problem* (3.9)–(3.10) *equals to zero.*

This lemma is a particular case of Theorem 2.1.

If $v = 0$ in $\mathbb{R}^3$, then $u_j = e^{ik\alpha \cdot x}$ cannot satisfy the boundary condition $\Gamma u_j = 0$ on S. Indeed, if $\Gamma u = u$, then $e^{ik\alpha \cdot s} \neq 0$ on S. If $\Gamma u = u_N$, then $ike^{ik\alpha \cdot s}\alpha \cdot N_s \neq 0$ for some $s \in S$. If $\Gamma u = u_N + \zeta(s)u$, then $ike^{ik\alpha \cdot s}\alpha \cdot N_s + \zeta(s)e^{ik\alpha \cdot s} \neq 0$ for some $s \in S$.

This contradiction proves that Case (a) cannot occur, so $S_1 = S_2$. The boundary condition on S is uniquely determined since $u(x, \alpha, k)$ is uniquely determined in D' by the scattering data.

Case (b).

As in Case (a) we prove that Equation (3.8) holds, where $D'_{12} = D'_1$ in Case (b).

Since v_2 is analytic in D'_2, the function $v_1 = v_2 = v$ continues analytically into $D_1 \setminus D_2$ as the solution to the equation

$$\left(\nabla^2 + k^2\right) v = 0 \quad \text{in } D'_2. \tag{3.11}$$

Therefore, $u = u_0 + v, u_0 := e^{ik\alpha \cdot x}$ satisfies the boundary conditions

$$\begin{aligned}
\Gamma_1 u &= 0 \quad \text{on} \quad S_1, &\tag{3.12}\\
\Gamma_2 u &= 0 \quad \text{on} \quad S_2, &\tag{3.13}
\end{aligned}$$

and the equation

$$\left(\nabla^2 + k^2\right) u = 0 \quad \text{in} \quad D'_2. \tag{3.14}$$

Let

$$g = g(x, y) := \frac{e^{ik|x-y|}}{4\pi|x - y|}. \tag{3.15}$$

By Green's formula one obtains

$$u(x, \alpha, k) = u_0(x, \alpha, k) + \int_S (ug_N - gu_N)ds, \tag{3.16}$$

where S can be S_1 or S_2. If $S = S_1$ then $x \in D'_1$ and if $S = S_2$ then $x \in D'_2$.

In the domain $D_1 \setminus D_2$ one derives using Green's formula the relation

$$u := u(x, \alpha, k) = \int_{S_1} (ug_N - gu_N)ds - \int_{S_2} (ug_N - gu_N)ds. \tag{3.17}$$

Note that N is the unit normal to S_j pointing out of D_j.

From formula (3.16) one gets

$$I_1 := \int_{S_1} (ug_N(x, s) - g(x, s)u_N)ds = \int_{S_2} (ug_N(x, s) - g(x, s)u_N)ds := I_2, \quad x \in D'_1. \tag{3.18}$$

Since I_2 is analytic in D_2' and I_1 is analytic in D_1' and they are equal in D_1' by Equation (3.18), it follows that I_1 can be continued analytically to $D_1 \setminus D_2$. Therefore, equalities (3.18) remain valid for $x \in D_2'$. This and formula (3.17) imply

$$u(x, \alpha, k) = 0 \quad \text{in} \quad D_1 \setminus D_2. \tag{3.19}$$

Since $u(x, \alpha, k)$ is analytic in D_2' and vanishes on the open subset of D_2' according to formula (3.19), it follows that $u(x, \alpha, k) = 0$ in D_2'. This is a contradiction since

$$\lim_{|x| \to \infty} |u(x, \alpha, k)| = 1. \tag{3.20}$$

This contradiction proves that $S_1 = S_2 := S$.

If $S_1 = S_2 := S$, then the boundary condition on S is uniquely determined by the scattering data.

Indeed, as we have already proved, the scattering solution $u(x, \alpha, k) := u_1(x, \alpha, k) = u_2(x, \alpha, k)$ is uniquely defined in $D' := D_1' = D_2'$.

Therefore, the quantity $\frac{u_N}{u}|_S$ is uniquely determined since, by elliptic regularity, $u(x, \alpha, k)$ is smooth up to the boundary S. If $\zeta(s) = 0$, then one has the Neumann condition on S. If $u = 0$ on S, that is, $\zeta(s) = \infty$ on S, then one has the Dirichlet condition on S. If $\zeta(s)$ is a continuous function, then one has the impedance boundary condition on S.

Our argument is valid for any choices of the boundary conditions on S_1 and S_2.

The conclusion is: Case (b) cannot occur.

Remark 3.5   There are important new features in our argument.

First, our argument is valid for the scattering data $A(\beta, k_0)$, that is, the non-over-determined data $A(\beta)$.

Second, Cases (a) and (b) were not discussed by other authors.

Third, it follows from our arguments that the following important result (formulated in Chapter 1 as Theorem 1.4) holds.

Lemma 3.6   *The scattering solution $u(x, \alpha, k)$ cannot have a closed surface of zeros different from S, the surface of the obstacle, if $u(s, \alpha, k) = 0$ for all $s \in S$.*

Case (c).

As in Case (a) we prove that the function w, defined in (3.1), satisfies relation (3.7). Equation (3.9) holds in D_{12}' and relation (3.10) holds. By Lemma 2.2 it follows that $w = 0$ in D_{12}', that is, relation (3.8) holds.

Since v_j, $j = 1, 2$, is analytic in D_j' and Equation (3.8) holds, one concludes that v admits unique analytic continuation as solution to the Helmholtz Equation (3.9) in $(D^{12})' := (\mathcal{D})'$.

Let us denote $D^{12} := \mathcal{D}_2$ and $D_{12} := \mathcal{D}_1$. Then $\mathcal{D}_2 \subseteq \mathcal{D}_1$, and we can use the argument from Case (b). This argument proves that $S_1 = S_2 := S$, and the boundary condition on S is uniquely determined.

In Case (b) the boundaries S_1 and S_2 were smooth. The boundaries S_{12} and S^{12} are not, in general, smooth if S_1 and S_2 intersect. However, formula (3.16) is valid since S_{12} and S^{12} are the unions of smooth pieces of surfaces. The role of S_2 now is played by S^{12} and the role of S_1 is played by S_{12}. The integrand in (3.16) is smooth up to the boundaries by the regularity properties of the solutions to the Helmholtz equation. Therefore, our proof in Case (b) remains valid with S^{12} and S_{12} replacing S_2 and S_1, respectively.

Theorem 3.3 is proved. □

Remark 3.7   Our proof of Theorem 3.3 is valid for non-over-determined scattering data $A(\beta) := A(\beta, k_0)$.

Let us sketch the idea of an earlier proof of the inverse obstacle scattering problem with the data $A(\beta, \alpha)$, see [5]. Only Case (c) was considered there. More details are given in Section 3.3.

Let us recall, see the beginning of the proof of Theorem 3.3, that $D_4 \subset D_1$ is a connected component of $D_3 = D_{12} \setminus D^{12}$.

If (3.8) holds, then v can be analytically continued in D_4 as a solution to the Helmholtz equation since v_2 is analytic in D_4 and $v = v_2$ in D'_{12}. Thus,

$$(\nabla^2 + k^2)v = 0 \quad \text{in } D_4, \tag{3.21}$$

$$\Gamma v = 0 \quad \text{on } \partial D_4, \tag{3.22}$$

where $\Gamma v = v$, that is, the Dirichlet boundary condition was assumed. The Dirichlet Laplacian in any bounded domain has a discrete spectrum. Since (3.21)–(3.22) holds for any $k \in (a, b), 0 \leq a < b, v \in L^2(D_4)$ for each of these k, one has a contradiction. This contradiction proves that $S_1 = S_2$.

It is known that the spectrum of Neumann Laplacian in domains with some non-smooth boundaries may be not discrete.

Therefore, for several decades there was a belief that the above argument cannot be used when the Neumann boundary condition on S was assumed. In [9] the above argument was slightly modified to overcome the difficulty with the spectrum. Namely, it is known that $L^2(D)$ is a separable Hilbert space, that is, it has a countable dense set. If (3.21)–(3.22) holds for every $k \in (a, b)$, then there exists a continuum of orthonormal vectors in $L^2(D_4)$, which contradicts to separability of $L^2(D_4)$. Indeed, if $k_1 \neq k_2$, then the usual argument yields

$$\int_D u(x, \alpha, k_1)\overline{u(x, \alpha, k_2)}dx = 0. \tag{3.23}$$

Thus, the reference to the fact that $L^2(D_4)$ is separable can replace the reference to the discreteness of the spectrum of the Laplacian. This argument, however, requires the scattering data to be over-determined.

3.3 UNIQUENESS OF THE SOLUTION TO THE INVERSE OBSTACLE SCATTERING PROBLEM WITH FIXED-ENERGY DATA

In this section we study the inverse obstacle scattering problem with the over-determined scattering data $A(\beta, \alpha, k) := A(\beta, \alpha)$, the fixed-energy data.

The first proof of the uniqueness of the solution to the above problem was given in [5]. The proof yielded Equations (3.21)–(3.22) in Case (c). To derive a contradiction in the case when $S_1 \neq S_2$ choose m values α_j of vector α, $1 \leq j \leq m$, such that the set $\{e^{ik\alpha_j \cdot x}\}_{j=1}^m$ is linearly independent in $L^2(D_4)$, where m is as large as one wishes. This choice of α_j is possible, see [5], and it implies that the set $\{u(x, \alpha_j, k_0)\}_{j=1}^m$ is linearly independent. Consequently, we have a contradiction: at a fixed $k > 0$ the dimension of the eigenspace of the Dirichlet Laplacian is bounded, say $\leq m_0$, while m can be larger than m_0.

The argument in [5] has used the assumption that the eigenspace of the Laplacian is finite-dimensional. For finite domains with non-smooth boundaries this is known to be true for the Dirichlet Laplacians. For domains with smooth boundaries this is known for any of the three types of boundary conditions: the D (Dirichlet), the N (Neumann), and the I(Impedance) boundary conditions.

The proof, the author gave for Theorem 3.3, assumes the smoothness of S_1 and S_2, but S_{12} and S^{12} are not smooth, in general. However, the integral formulas we use in the proof of Theorem 3.3 are valid, the non-smooth surfaces S_{12} and S^{12} are unions of smooth surfaces and the integrands are smooth up to these surfaces.

3.4 UNIQUENESS OF THE SOLUTION TO INVERSE OBSTACLE SCATTERING PROBLEM WITH NON-OVER-DETERMINED DATA

The basic result, Theorem 3.3, was already proved. It says that S and the boundary condition on S are uniquely determined by the data $A(\beta) := A(\beta, \alpha_0, k_0)$.

Let us discuss some consequences of this uniqueness result.

Theorem 3.8   *The data* $A(\beta) := A(\beta, \alpha_0, k_0)$ *determines uniquely* $A(\beta, \alpha, k)$.

Proof. Since $A(\beta) = A(\beta, \alpha_0, k_0)$ determines uniquely S and the boundary condition on S, it determines uniquely $A(\beta, \alpha, k)$.

Theorem 3.8 is proved.   □

Let us derive a useful formula which gives a relation between scattering amplitudes for two obstacles. This theorem was proved by the author for the Schrödinger operators; see [9].

Theorem 3.9　*One has*

$$4\pi\left[A_1(\beta,\alpha,k) - A_2(\beta,\alpha,k)\right] =$$
$$= \int_{S_{12}} \left[u_1(s,\alpha,k)u_{2N}(s,-\beta,k) - u_{1N}(s,\alpha,k)u_2(s,-\beta,k)\right] ds. \tag{3.24}$$

Proof. Let $k > 0$ be fixed. We drop the k-dependence for brevity. One has

$$\left(\nabla_x^2 + k^2\right)G_1(x,y) = -\delta(x-y), \tag{3.25}$$
$$\left(\nabla_x^2 + k^2\right)G_2(x,z) = -\delta(x-z). \tag{3.26}$$

Multiply the first equation by $G_2(x,z)$, the second equation by $G_1(x,z)$, subtract from the resulting first equation the second one, integrate over $B_R \setminus D_{12}$, and use Green's formula. The result is

$$G_1(x,y) - G_2(x,y) = \int_{S_{12}} \left[G_1(s,x)G_{2N}(s,y) - G_{1N}(s,x)G_2(s,y)\right] ds, \tag{3.27}$$

where we have used the radiation condition for G_1 and G_2, took $R \to \infty$ and replaced z, y by x, y in the result. Let us use Lemma 2.7. Take $|y| \to \infty$, $\frac{y}{|y|} = \alpha$ in (3.27) and obtain by Lemma 2.7 the following relation

$$u_1(x,\alpha,k) - u_2(x,\alpha,k) = \int_{S_{12}} \left[G_1(s,x)u_{2N}(s,\alpha,k) - G_{1N}(s,x)u_2(s,\alpha,k)\right] ds. \tag{3.28}$$

Remember that

$$u_j(x,\alpha,k) = e^{ik\alpha\cdot x} + \frac{e^{ik|x|}}{|x|}A(\beta,\alpha,k) + o\left(\frac{1}{|x|}\right), \quad |x| \to \infty, \quad \frac{x}{|x|} = \beta. \tag{3.29}$$

Let $|x| \to \infty$, $\frac{x}{|x|} = -\beta$ in (3.28). Use Lemma 2.7 to get:

$$A_1(\beta,\alpha,k) - A_2(\beta,\alpha,k) =$$
$$= \frac{1}{4\pi}\int_{S_{12}} \left[u_1(s,-\beta,k)u_{2N}(s,\alpha,k) - u_{1N}(s,-\beta,k)u_2(s,\alpha,k)\right] ds. \tag{3.30}$$

This formula is equivalent to formula (3.24) due to Lemma 2.9, that is, due to formula (2.55). Theorem 3.9 is proved.　□

Lemma 3.10   *Let*

$$\left(\nabla^2 + k^2\right) G(x, y, k) = -\delta(x - y) \quad in \quad D',$$ (3.31)

$$G|_S = 0,$$ (3.32)

$$\frac{\partial G(x, y, k)}{\partial |x|} - ikG(x, y, k) = o\left(\frac{1}{|x|}\right), \quad |x| \to \infty,$$ (3.33)

where (3.33) is valid for $|y| \leq R$ uniformly with respect to the directions $\frac{x}{|x|}$. Let $t \in S$. Then

$$\lim_{x \to t} G_N(x, s, k) = \delta(s - t),$$ (3.34)

where $\delta(s - t)$ is the delta-function concentrated on S.

Proof. Let $f \in C(S)$ be arbitrary. Let

$$\left(\nabla^2 + k^2\right) \psi = 0 \quad in \quad D',$$ (3.35)

$$\psi|_S = f,$$ (3.36)

$$\frac{\partial \psi}{\partial |x|} - ik\psi = o\left(\frac{1}{|x|}\right), \quad |x| \to \infty.$$ (3.37)

This problem has a unique solution which can be calculated by Green's formula

$$\psi(x) = \int_S (\psi G_N - \psi_N G) ds = \int_S G_N(x, s) f(s) ds, \quad x \in D',$$ (3.38)

where condition (3.32) was used.

Let $x \to t \in S$. Then $\psi(x) \to f(t)$ and from (3.36) and (3.38) the relation (3.34) follows. Lemma 3.10 is proved. □

Let D_1 and D_2 be two bounded domains, $A_j(\beta, \alpha, k)$ are corresponding scattering amplitudes, $G_j(x, y, k)$, and the boundary conditions on S_j are either the Dirichlet, the Neumann, or the impedance type.

Theorem 3.11   *If $A_1(\beta, \alpha_0, k_0) = A_2(\beta, \alpha_0, k_0)$, then $A_1(\beta, \alpha, k) = A_2(\beta, \alpha, k)$ and $G_1(x, y, k) = G_2(x, y, k)$.*

Proof. If $A_1(\beta, \alpha_0, k_0) = A_2(\beta, \alpha_0, k_0)$, then by Theorem 3.3, S and the boundary condition on S are uniquely determined. Consequently, $G_1(x, y, k) = G_2(x, y, k)$. Therefore, $A_1(\beta, \alpha, k) = A_2(\beta, \alpha, k)$.

Theorem 3.11 is proved. □

3.5   NUMERICAL SOLUTION OF THE INVERSE OBSTACLE SCATTERING PROBLEM WITH NON-OVER-DETERMINED DATA

In this section we assume that the obstacle is two-dimensional, S is a curve with the equation

$$r = f(\phi). \tag{3.39}$$

The Green's function of the Helmholtz equation in $\mathbb{R}^2$ is

$$g(x, y) = \frac{i}{4} H_0^{(1)}(k|x - y|), \tag{3.40}$$

$$(\nabla^2 + k^2) g(x, y) = -\delta(x - y). \tag{3.41}$$

Consider the scattering problem

$$(\nabla^2 + k^2)u = 0 \quad \text{in} \quad D' \subset \mathbb{R}^2, \tag{3.42}$$

$$u|_S = 0, \tag{3.43}$$

$$u = e^{ik\alpha \cdot x} + v, \tag{3.44}$$

$$v_r - ikv = o\left(\frac{1}{r}\right), \quad r = |x| \to \infty. \tag{3.45}$$

Let us use Green's formula

$$u(x) = \int_{S_R} (gu_N - ug_N)ds - \int_S (gu_N - ug_N)ds. \tag{3.46}$$

As $R \to \infty$ the first integral tends to u_0. Let us explain this. One has:

$$\lim_{R \to \infty} \left(\int_{S_R} (gu_{0N} - u_0 g_N)ds + \int_{S_R} (gv_N - vg_N)ds \right) = u_0(x) := u_0(x, \alpha, k). \tag{3.47}$$

The limit of the second integral in (3.47) is zero by the radiation condition, while the limit of the first integral in (3.47) is u_0 by the Green's formula. For any $x \in B_R$ the first integral equals to u_0 by Green's formula.

Thus, formula (3.46), as $R \to \infty$, can be written as

$$u(x) = u_0(x) - \int_S g(x, s)h(s)ds, \quad h := u_N|_S. \tag{3.48}$$

Using the known formula for the normal derivative of the potential of single layer (see, for example, [9], p. 119, formula 2.2.11) one gets:

$$h = -\mathcal{A}h + 2u_{0N}, \tag{3.49}$$

where

$$Ah = 2 \int_S g_N(s,t)h(t)dt. \tag{3.50}$$

Denote

$$\|f\| = \left(\int_S |f(\phi)|^2 d\phi \right)^{1/2}, \quad \|f\|^2 := N(f). \tag{3.51}$$

Consider the following minimization problem:

$$F(f,h) = N\left(A(\beta) + \int_S e^{-ik\beta \cdot s}h(s)ds \right) + N(h + Ah - 2u_{0N}) = \min. \tag{3.52}$$

The significant advantage of the functional $F := F(f,h)$ is the following one.

If $A(\beta)$ are the exact data, then the functional F attains its unique minimum, equal to zero, on the unique pair $\{f, h\}$, where $r = f(\phi)$ is the equation of S and h is the unique solution to Equation (3.49) *in which the operator A is uniquely defined if f is known.*

Here $A(\beta)$ are the values of the scattering amplitude given by measurements. These values can be theoretically calculated from formula (3.48):

$$A(\beta) = c \int_S e^{-ik\beta \cdot s}h(s)ds, \tag{3.53}$$

$$c = -\frac{i}{4}\sqrt{\frac{2}{\pi k}}e^{-i\frac{\pi}{4}}. \tag{3.54}$$

The minimization in (3.52) is over $f(\phi)$ and $h(s)$. The following formulas allow one to write (3.52) so that the dependence on f is explicit. The radius vector of the boundary is

$$s = f(\phi)\cos\phi e_1 + f(\phi)\sin\phi e_2, \tag{3.55}$$

where $\{e_1, e_2\}$ is an orthonormal basis of $\mathbb{R}^2$.

The unit normal to S is

$$N = \frac{-[f'(\phi)\sin\phi + f(\phi)\cos\phi]e_1 + [f'(\phi)\cos\phi - f(\phi)\sin\phi]e_2}{[(f'^2(\phi))^2 + f^2(\phi)]^{1/2}}. \tag{3.56}$$

One has

$$ds = \left[(f'(\phi))^2 + f^2(\phi) \right]^{1/2} d\phi, \tag{3.57}$$

$$f'(\phi) \approx \frac{f(\phi + \Delta\phi) - f(\phi - \Delta\phi)}{2\Delta\phi}, \tag{3.58}$$

$$h(s) = h(f(\phi)\cos\phi e_1 + f(\phi)\sin\phi e_2), \tag{3.59}$$

$$g_N(s,t) = \nabla_x g(x,t)|_{x=s} \cdot N. \tag{3.60}$$

Formulas (3.54)–(3.60) allow one to calculate functional (3.52) and consider this functional as the functional of f and h. If S is known, that is f is known, then h can be uniquely found from Equation (3.49).

One can write the functional (3.52) in a discretized form for numerical purposes. Here we want to concentrate on the basic ideas of the numerical method.

Let

$$M = \{f : ||f||_2 = ||f|| + ||f'|| + ||f''|| \leq m\}. \tag{3.61}$$

The set M is a compact set in the space $C([0, 2\pi])$ of continuous functions on S with the usual norm. If the equation of S in polar coordinates is $r = f(\phi)$ and $f \in M$, then

$$F(f, h) = 0, \tag{3.62}$$

for exact data $A(\beta)$, because h is the unique solution to (3.49) for a given $f \in M$.

If (3.62) holds then (3.53) holds if the data $A(\beta)$ are exact.

By Theorem 3.3 the $A(\beta)$ determines S and, therefore, $f = f(\phi)$ uniquely. Therefore, for exact data $A(\beta)$ the functional F in (3.52) has a unique minimum, its minimum is equal to zero and it is attained on the function f, such that $r = f(\phi)$ is the equation of S, and h solves Equation (3.49).

If the data $A_\epsilon(\phi)$ is not exact, then

$$||A_\epsilon(\phi) - A(\phi)|| \leq \epsilon, \tag{3.63}$$

where $A(\beta)$ are the exact data.

Let

$$\lim_{n \to \infty} ||f_n - f||_2 = 0, \tag{3.64}$$

where we assume that $f \in M$ defines the exact S.

Let us formulate our result.

Theorem 3.12 *If the data $A(\beta)$ are exact and*

$$\lim_{n \to \infty} F(f_n, h_n) = 0, \tag{3.65}$$

then

$$\lim_{n \to \infty} ||f_n - f|| = 0, \tag{3.66}$$

and $r = f(\phi)$ is the equation of S.

Proof. From (3.65) it follows that

$$\lim_{n \to \infty} ||A(\beta) + c \int_{S_n} e^{-ik\beta \cdot s} h_n(s) ds|| = 0, \tag{3.67}$$

where S_n is the surface defined by the equation $r = f_n(\phi)$. Since $f_n \in M$ there is a subsequence, denoted again f_n, such that (3.66) holds. If (3.66) holds, then the operator A_n in the Equation (3.49), with A_n in the place of A, converges to A, corresponding to S, that is, to f.

Since $A(\beta)$ determines uniquely S, that is, f is determined uniquely, Theorem 3.12 is proved. $\square$

Suppose that the data are not exact and the inequality (3.63) holds. If $||A_\epsilon(\phi) - A(\phi)|| = \epsilon$, then

$$\epsilon \leq F(f, h). \tag{3.68}$$

If (3.63) and (3.65) hold for exact data, then for noisy data one has

$$\lim_{n \to \infty} F(f_n, h_n) \leq \epsilon^2. \tag{3.69}$$

Therefore, from the formula

$$F(f_n, h_n) = ||A_\epsilon(\beta) + c \int_{S_n} e^{-ik\beta \cdot s} h_n(s) ds||^2 + ||h_n + A_n h_n - 2u_{0N}||^2 \leq \epsilon^2, \quad n \to \infty, \tag{3.70}$$

it follows that

$$||A_\epsilon(\beta) + c \int_{S_n} e^{-ik\beta \cdot s} h_n(s) ds||^2$$

$$\leq 2||A_\epsilon(\beta) - A(\beta)||^2 + 2||A(\beta) + c \int_{S_n} e^{-ik\beta \cdot s} h_n(s) ds||^2 \tag{3.71}$$

$$\leq 2\epsilon^2 + 2||A(\beta) + c \int_{S_n} e^{-ik\beta \cdot s} h_n(s) ds||^2.$$

Since for exact data there exists a sequence $\{f_n, h_n\}$ such that (3.65) holds, for this sequence one has

$$F(f_n, h_n) \leq 3\epsilon^2, \quad n > n(\epsilon). \tag{3.72}$$

Thus, the equation $r = f_n(\phi), n > n(\epsilon)$, one may consider as an approximate equation of S for noisy data.

There were no attempts, to our knowledge, to solve numerically the inverse obstacle scattering data with non-over-determined data.

Critical remarks about attempts to solve this problem with over-determined data can be found in [18, pp. 245–253]. There one can find references to the papers and books dealing with this problem.

The arguments in this section shows that the solution we constructed for inverse obstacle scattering problem is stable in the following sense: the equation of the boundary $r = f_n(\phi)$ satisfies the relation $||f_n - f|| < \epsilon$ for $n > n(\epsilon)$. But small perturbations $A_\epsilon(\phi)$ of the exact data $A(\phi)$ may be functions that are not scattering amplitudes for any bounded obstacle.

APPENDIX A

Existence and Uniqueness of the Scattering Solutions in the Exterior of Rough Domains

This appendix is essentially paper [25] with minor changes.

A.1 INTRODUCTION

In this appendix we study the scattering problem in the exterior of a rough bounded domain. This problem was investigated in [6] and [24], where it was assumed that the potential had compact support. The goal of this appendix is to relax this hypothesis and to give a simple and general method of proof which is simpler than the earlier known. The assumptions on the coefficients of the differential operator are also relaxed. We prove the results assuming that the potential decays at infinity at the power rate which depends on the space dimension. We discuss the three-dimensional case ($n = 3$), but the arguments are similar in the n-dimensional case. By this reason in many places we keep n in the formulation of the results and assumptions. Estimates (A.40), (A.41), (A.43), (A.44), and (A.58) are given for $n = 3$.

   The assumptions on the smoothness of the domain are minimal and include all the previously studied cases and probably all of the cases of interest in applications.

   This appendix is organized as follows. In the rest of the introduction we introduce some notations, the function spaces we work with, and give the statement of the scattering problem. In Section A.2 we prove the uniqueness of the solution of the scattering problem. In Section A.3 we prove the existence of the solution assuming that the potential has compact support. In Section A.4 we assume a power decay rate of the potential and prove the existence of the scattering solution.

   Our results are stated and proved for the broader class of domains than in the literature, see, e.g., [27]. In particular, Lipschitz domains form a proper subset in the class of domains we study. An inverse obstacle scattering problem in a class of rough domains was studied in [6].

   The basic ideas of our method are simple: first, we prove that the operator of the problem under consideration is selfadjoint. This is done by using the fact that every closed symmetric densely defined semibounded from below quadratic form on a Hilbert space defines a unique selfadjoint operator whose domain is dense in the domain of the quadratic form. The assumptions (A.4) and (A.5) concerning the roughness of the domain guarantee that the corresponding quadratic form is semibounded from below and closed. The case of the Dirichlet boundary condition is the simplest one, and in this appendix we mostly deal with the impedance (Robin) boundary condition. In the simplest case of the Dirichlet boundary condition no assumptions are needed on the roughness of the obstacle: any compact domain is admissible. The reason is simple: for the compactness of the embedding $H_0^1(D_R) \to L^2(D_R)$ no assumptions on D, except boundedness of D are needed, while for the compactness of the embedding $H^1(D_R) \to L^2(D_R)$ (see next section for the notations and condition (A.4)) some assumptions concerning the roughness of the boundary of D are necessary. Necessary and sufficient conditions for the compactness of the above embedding operator are known [4]. For the Robin boundary condition in addition to (A.4) assumption (A.5) is needed.

Second, we prove that the solution to the problem with complex spectral parameter $k^2 + i\epsilon$, $\epsilon > 0$, converges, as $\epsilon \to 0$, to the solution of the scattering problem. The convergence holds in suitable local and global norms.

The underlying idea is to use the Fredholm property of the problem in a properly chosen pair of normed spaces.

A.1.1 NOTATIONS AND ASSUMPTIONS

The following are the notations used in this appendix.

By $D \subset R^n$, we denote a bounded domain, D' stands for its complement, $D' := R^n \setminus D$, $D_R := D' \cap B_R$, B_R denotes a ball with radius R such that $B_R \supset D$, and B'_R is the complement of B_R in R^n.

By $a_{ij}(x)$, $x \in D'$, we denote the elements of a real-valued symmetric matrix satisfying the following ellipticity condition:

$$\exists c, C > 0 \text{ such that} \qquad c|t|^2 \le a_{ij}(x)t_i\bar{t}_j \le C|t|^2 \qquad \forall t \in C^n, \forall x \in D'. \qquad (A.1)$$

In Equation (A.1) and below the summation over the repeated indices is understood. One has:

$$a_{ij}(x) = \delta_{ij} \text{ when } |x| > R, \qquad (A.2)$$

and the coefficients $a_{ij}(x)$ are assumed Lipschitz continuous. This assumption implies the validity of the unique continuation principle for the solutions to elliptic Equation (A.8) below. By an over-line the complex conjugate is denoted. Let

$$lu := -\partial_j \left(a_{ij}(x)\partial_i u \right).$$

Assume

$$q(x) = \overline{q(x)}, \quad q(x) \in L_{\text{loc}}^p(R^n), \quad p > \frac{n}{2}, \quad q(x) \in L^\infty(\tilde{S}),$$

$$|q(x)| \le \frac{c}{(1 + |x|^2)^{s/2}}, \quad |x| > R, \ s > n, \ n \ge 3, \qquad (A.3)$$

where $\tilde{S} \subset D'$ is a neighborhood of the boundary $S := \partial D$. The bar stands for complex conjugate. The assumption $s > n$ is weakened in some of the statements of this appendix: for example, in Proposition A.9 below it is assumed that $s > 1$. However, the assumption

$$s > n, n = 3$$

is used in Section A.4.

In [1] the scattering theory is developed under the assumptions that $a_{ij} = \delta_{ij}$, the obstacle is absent, and $q(x)$ satisfies less restrictive assumptions than (A.3)—essentially it is assumed that $s > 1$. Therefore, it is likely that the results of our appendix can be obtained under this weaker assumption. However, additional technical work is needed to obtain such results. We are emphasizing the methodology of handling rough boundaries in this appendix. Discussion of the wider class of potentials would lead us astray.

Let

$$Lu := lu + q(x)u.$$

Let $S := \partial D$. Our main assumptions concerning the smoothness of D are:

$$\text{The embedding} \quad i : H^1(D_R) \to L^2(D_R) \quad \text{is compact,} \qquad (A.4)$$

and

$$\text{The embedding} \quad r : H^1(D_R) \to L^2(S) \quad \text{is compact,} \qquad (A.5)$$

where in the definition of $L^2(S)$ the integration over S is understood with respect to the $n - 1$-dimensional Hausdorff measure.

In particular, assumption (A.5) implies that the $n - 1$-dimensional Hausdorff measure of S is finite, that D has a finite perimeter (see [4], p. 296), and that the Gauss–Green formula holds in D for functions in $BV(D)$ space, which consists of functions $u \in L^1_{loc}(D)$ such that ∇u, understood in the sense of distribution theory, is a signed measure (a charge) in D.

Let

$$u_N := \partial_N u := a_{ij}(x)\partial_i u N_j, \quad x \in S,$$

where N_j is the j-*th* component of the normal to S pointing into D.

The definition of the normal for rough domains is not discussed here because in our formulation of the scattering problem (see (A.11)–(A.12)) the notion of normal is not used. One can find the definition of the normal in the sense of Federer in [4], p. 303.

In this appendix we do not define the class of rough domains explicitly, but isolate property (A.4) or properties (A.4) and (A.5) as defining properties. In [4], necessary and sufficient conditions on S, the boundary of D, are given for these properties to hold.

In formulas below in which the surface integrals appear, e.g., (A.11), (A.23), (A.24), etc., the integration measure ds is the $n - 1$-dimensional Hausdorff measure defined, for example in [4], p. 37. In [2], it is proved that for rectifiable surfaces (the class of these surfaces is much larger than the class of Lipschitz surfaces) the $n - 1$-dimensional Hausdorff measure is equivalent to the measure generated by the elements of the surface area.

Given an $L^\infty(S)$ real-valued function $\sigma(s)$, $s \in S$, one defines the boundary condition operator Γ^R, corresponding to Robin or impedance boundary condition:

$$\Gamma^R u(s) := u_N(s) + \sigma(s)u(s), \quad s \in S \quad \sigma \geq 0.$$

Introduce the following Dirichlet and Neumann boundary conditions operators:

$$\begin{aligned} \Gamma^D u(s) &:= u(s), \quad s \in S, \\ \Gamma^N u(s) &:= u_N(s), \quad s \in S. \end{aligned}$$

Let $u_0 := \exp(ik\alpha \cdot x)$, $\alpha \in S^{n-1}$, where S^{n-1} is the unit sphere in $\mathbf{R}^n$, $k > 0$, and

$$v := u - u_0.$$

A.1.2   FUNCTION SPACES

Let

$$(u, v) := \int_{D'} u\bar{v} dx. \tag{A.6}$$

On $H^1(D')$ define the following bilinear form, $[\cdot, \cdot] : H^1(D') \times H^1(D') \to C$:

$$[u, v] := \int_{D'} a_{ij} \partial_i u \partial_j \bar{v} dx + (u, v). \tag{A.7}$$

We use the following spaces.

Definition A.1   The space $L^2_{\text{loc}}(D')$ is the space of functions f such that, $\forall r > R$, $f \in L^2(D_r)$.

Definition A.2   The space $H^l_{\text{loc}}(D')$ is the space of functions f such that, $\forall r > R$, $f \in H^l(D_r)$, where $H^l(D_r)$ is the Sobolev space.

Definition A.3   The space $H^1_0(D')$ is the closure in $H^1(D')$ norm of the space of functions $f \in H^1(D')$ vanishing near S. The set of functions which belong to $H^1_0(D_r)$ for any $r > R$ is denoted by $H^1_{0loc}(D')$.

Definition A.4   The space $\tilde{H}^1(D')$ is the space of functions $f \in H^1(D')$ vanishing near infinity.

Definition A.5 The space $\tilde{H}_0^1(D')$ is the space of functions $f \in H_0^1(D')$ vanishing near infinity.

Definition A.6 The space $H_s^\ell := H_s^\ell(D')$, $\ell = 0, 1$, is the space of functions $f \in H_{\mathrm{loc}}^\ell(D')$ with finite weighted $L^2(D')$-norm, for example the norm in H_s^ℓ with $\ell = 1$ is defined as follows:

$$\|f\|_{1,s}^2 = \int_{D'} \left(1 + |x|^2\right)^{s/2} \left(|f|^2 + |\nabla f|^2\right) dx < \infty.$$

A.1.3 STATEMENT OF THE PROBLEM

We study the following problem:

$$Lu - k^2 u = 0 \quad \text{in } D', \tag{A.8}$$

$$\Gamma^X u = 0 \quad \text{in } S, \tag{A.9}$$

$$\lim_{r \to \infty} \int_{|x|=r} |v_r - ikv|^2 ds = 0, \tag{A.10}$$

where $v = u - u_0$, $u_0 = e^{ik\alpha \cdot x}$, $\alpha \in S^{n-1}$ is given, $k > 0$ is fixed, and $X = R$ (the Robin boundary condition), $X = N$ (the Neumann boundary condition), or $X = D$ (the Dirichlet boundary condition). We discuss mostly the Robin boundary condition. The other cases can be treated similarly. The Neumann boundary condition is a particular case of the Robin boundary condition ($\sigma(s) \equiv 0$). The Dirichlet boundary condition is the simplest:

It does not require any smoothness assumptions concerning the boundary S; only the boundedness of D is required.

For the Neumann boundary condition only assumption (A.4) is needed.

For the Robin boundary condition both assumptions (A.4) and (A.5) are used.

The function v is the scattered field and u_0 is the incident field, a plane wave.

A.1.4 THE WEAK FORMULATION OF THE SCATTERING PROBLEM

Here we introduce the weak formulation of problem (A.8)–(A.10).

Definition A.7 We say that $u \in H_{loc}^1(D') \cap H_{-s}^1(D')$, $s > 1$, is a weak solution of the scattering problem (A.8)–(A.10) with $X = R$ (the Robin boundary condition) if

$$\int_{D'} \left[a_{ij} \partial_i u \partial_j \bar{\varphi} + (q - k^2) u \bar{\varphi}\right] dx + \int_S \sigma u \bar{\varphi} ds = 0 \quad \forall \varphi \in \tilde{H}^1(D'), \tag{A.11}$$

$$\lim_{r \to \infty} \int_{|x|=r} |v_r - ikv|^2 ds = 0, \quad \text{where} \quad v = u - u_0. \tag{A.12}$$

Definition A.8   We say that $u \in H^1_{0\,loc}(D') \cap H^1_{-s}(D')$, $s > 1$, is a weak solution of the scattering problem (A.8)–(A.10) with $X = D$ (the Dirichlet boundary condition) if

$$\int_{D'} \left[a_{ij} \partial_i u \partial_j \bar{\varphi} + \left(q - k^2 \right) u \bar{\varphi} \right] dx = 0 \quad \forall \varphi \in \tilde{H}^1_0 (D'), \tag{A.13}$$

$$\lim_{r \to \infty} \int_{|x|=r} |v_r - ikv|^2 ds = 0, \quad \text{where} \quad v = u - u_0. \tag{A.14}$$

A.2   UNIQUENESS THEOREM

In this section we prove uniqueness of the solution to problem (A.8)–(A.10). We first state the following known results.

Proposition A.9   *([29], p. 227) Suppose u is a solution of Equation (A.8) vanishing on an open subset of D'. If $a_{ij}(x)$ are Lipschitz and $q(x) \in L^p_{loc}(R^n)$, $p > \frac{n}{2}, n \geq 3$, then $u \equiv 0$ in D'.*

The above Proposition is called the *unique continuation principle*.

Proposition A.10   *([5], p. 25) Suppose that (A.2) and (A.3) hold, $s > 1$ in (A.3) and $k > 0$. If $u \in H^1_{loc}(B'_R)$ and*

$$\int_{B'_R} \left[\partial_i u \partial_i \bar{\varphi} + \left(q(x) - k^2 \right) u \bar{\varphi} \right] dx = 0 \quad \forall \varphi \in \tilde{H}^1_0(B'_R), \tag{A.15}$$

$$\lim_{r \to \infty} \int_{|x|=r} |u|^2 ds = 0, \tag{A.16}$$

then $u \equiv 0$ outside B_R.

In [5, p. 25], it is assumed that $q(x) = 0$. In the general case the result follows from a theorem of T. Kato (see [3]) which says that any solution to Equation (A.11) in B'_R, which satisfies (A.16), vanishes in B'_R if $k > 0$ and $|x||q(x)| \to 0$ as $|x| \to \infty$. It is not necessary to assume q real-valued in Kato's theorem.

We now prove a lemma used in the proof of the uniqueness of the solution of the scattering problem.

Lemma A.11   *Suppose W satisfies (A.11). Then*

$$\int_{|x|=r} \left(\bar{W} W_r - W \bar{W}_r \right) ds = 0 , \quad r > R. \tag{A.17}$$

Proof. Take a cut-off function $h(r) \in C^\infty(R)$ such that $0 \le h \le 1$, $h(r) = 1$ when $r \le 1/2$, $h(r) = 0$ when $r \ge 3/2$, h is monotonically decreasing. In (A.11) take $\bar{\varphi} = \bar{W} h\left((|x| - r_0)/\delta\right)$, with $r_0 > R$, and $\delta > 0$. Then take the complex conjugate and subtract. Using (A.2), one gets:

$$\int_{D'} \left(\bar{W} \partial_i W - W \partial_i \bar{W} \right) \partial_i h \left((|x| - r_0)/\delta\right) dx = 0.$$

If one takes the limit $\delta \to 0$ in the above relation, one gets the desired result. Taking this limit, one uses the interior regularity of W: this regularity is a consequence of Equation (A.11) and of the known interior elliptic regularity results. $\qquad \square$

We now state the main result of this section.

Theorem A.12 *Suppose u_1 and u_2 are the weak solutions of the scattering problem with the Robin (or the Neumann, or the Dirichlet) boundary condition. Then $u_1 \equiv u_2$.*

Proof. We prove this Theorem in the case of the Robin boundary condition. The cases of the Neumann and the Dirichlet boundary conditions can be treated similarly.

Define $W := u_1 - u_2$. One has:

$$\int_{D'} \left[a_{ij} \partial_i W \partial_j \bar{\varphi} + \left(q - k^2\right) W \bar{\varphi} \right] dx + \int_S \sigma W \bar{\varphi} ds = 0 \quad \forall \varphi \in \tilde{H}^1\left(D'\right), \quad \text{(A.18)}$$

$$\lim_{r \to \infty} \int_{|x|=r} |W_r - ikW|^2 ds = 0. \qquad \text{(A.19)}$$

Equation (A.19) can be written as:

$$\lim_{r \to \infty} \int_{|x|=r} \left(|W_r|^2 + k^2 |W|^2\right) ds + \lim_{r \to \infty} ik \int_{|x|=r} \left(W_r \bar{W} - \bar{W}_r W\right) ds = 0.$$

Since W satisfies (A.18), Lemma A.11 applies, and the above relation implies

$$\lim_{r \to \infty} \int_{|x|=r} |W|^2 ds = 0,$$

which is condition (A.16) of Proposition A.10. Moreover, from (A.18) and (A.2) it follows that W satisfies condition (A.15). Therefore, Proposition A.10 implies $W = 0$ outside B_a. Then one applies Proposition A.9 and concludes that $W \equiv 0$ in D'.

Theorem A.12 is proved. $\qquad \square$

Having proved the uniqueness of the scattering solution, we now wish to prove the existence of this solution.

Let us reduce the problem to the one for a function which satisfies the radiation condition at infinity.

Take a cut-off function $\zeta \in C^\infty(R)$, such that $0 \le \zeta \le 1$, $\zeta \equiv 0$ in a neighborhood of D and $\zeta \equiv 1$ outside B_r for some $r > R$, and define

$$w := u - \zeta u_0.$$

If u solves the scattering problem (A.8)–(A.10), then w solves the following problem:

$$Lw - k^2 w = f := \left(L - k^2\right)(\zeta u_0), \tag{A.20}$$

$$\Gamma^X w = 0, \tag{A.21}$$

$$\lim_{r \to \infty} \int_{|x|=r} |w_r - ikw|^2 ds = 0. \tag{A.22}$$

In Section A.3 we prove that the above problem has a solution. Theorem A.12 implies that this solution is unique.

A.3   EXISTENCE OF THE SCATTERING SOLUTIONS FOR COMPACTLY SUPPORTED POTENTIALS

In this section we prove that the scattering problem (A.8)–(A.10) with a compactly supported potential, $q(x) = 0$ if $|x| > a$, has the unique solution. We look for a solution of the form $u = w + \zeta u_0$, where ζ is a cut-off function (defined above formula (A.20)) and w satisfies (A.20)–(A.22). Let us prove that problem (A.20)–(A.22) has a solution.

A.3.1   EXISTENCE FOR THE EQUATION WITH THE ABSORPTION

Consider the weak form of the scattering problem.

Let $\varepsilon > 0$. Find $w \in H^1_{loc}(D') \cap H^1_{-s}(D')$, $s > 1$, such that:

$$\int_{D'} \left[a_{ij} \partial_i w \partial_j \bar{\varphi} + \left(q - k^2 - i\varepsilon\right) w \bar{\varphi} \right] dx + \int_S \sigma w \bar{\varphi} ds = (f, \varphi) \quad \forall \varphi \in \tilde{H}^1(D'). \tag{A.23}$$

In the space $H^1(D') \subset L^2(D')$ consider the following bilinear form:

$$B_\gamma[u, v] := [u, v] + \gamma(u, v) + \int_S \sigma u \bar{v} dS + (qu, v) := [u, v]_\gamma + (qu, v), \tag{A.24}$$

where $[u, v]$ is defined in (A.7) and $\gamma > 0$ is chosen so large that the form $[u, u]_\gamma^{1/2}$ defines the norm equivalent to $H^1(D')$. We use here assumption (A.5), which implies that the boundary integral in (A.24) can be estimated by the term

$$c\|u\|_{H^1(D_R)} \|v\|_{H^1(D_R)}.$$

One could omit assumption (A.5) and change the space in which the solution is sought, to the space with the norm containing the additional term $\|u\|_{L^2(S)}$, where the measure used in the definition of $L^2(S)$ is the $n-1$-dimensional Hausdorff measure. Then one could use assumption (A.4), do not use assumption (A.5), and assume that S has finite perimeter (see [28]).

A set has finite perimeter if the gradient (in the sense of distribution theory) of the characteristic function of this set belongs to the space $BV(D)$. This space was mentioned below formula (A.5) (see also [28], p. 152).

If $X = D$ in (A.9), then the term $\int_S \sigma u \bar{v} dS$ is absent in the definition of $[u, v]_\gamma$ in (A.24) and assumptions (A.4) and (A.5) can be dropped.

If $X = N$ in (A.9), then the term $\int_S \sigma u \bar{v} dS$ is also absent in the definition of $[u, v]_\gamma$ in (A.24), assumption (A.4) is used and assumption (A.5) is not used.

If $X = R$ in (A.9), then both assumptions (A.4) and (A.5) are used.

Assumption (A.4) allows one to conclude that (A.5) implies (A.34) (see below) in the case of Neumann or Robin boundary conditions.

Assumption (A.5) allows one to conclude that (A.28) implies that the term $\int_S \sigma \psi_n \bar{\varphi} \, dS$ converges and can be estimated by the $H^1(D_R')$-norm of ψ_n for any test function φ in (A.36) (see below).

Assuming (A.1)–(A.5), one has the following.

Lemma A.13   *The form $B_\gamma[\cdot, \cdot]$ is continuous in $H^1(D') \times H^1(D')$ and for all sufficiently large $\gamma > 0$ there exist $\beta_j, j = 1, 2$, such that*

$$\beta_1 \|u\|_1^2 \leq B_\gamma[u, u] \leq \beta_2 \|u\|_1^2 \qquad \beta_1 > 0.$$

Proof. From (A.3), (A.4) and the result from [7] one gets:

$$\left| \int_{\tilde{S}} q|u|^2 dx \right| \leq \nu \|u\|_{H^1(\tilde{S})}^2 + C(\nu)\|u\|_{L^2(\tilde{S})}^2,$$

for any $\nu > 0$, however small. Denote $\tilde{D}_R := D_R \setminus \tilde{S}$, and choose $\tilde{S}$ such that the boundary of $\tilde{D}_R$ is smooth. One has

$$\left| \int_{\tilde{D}_R} q|u|^2 dx \right| \leq c(R)\|u\|_1^2$$

if (A.3) holds. Indeed, if $u \in H^1(\tilde{D}_R)$, then, by the Sobolev embedding theorem for domains with sufficiently smooth (say, Lipschitz) boundaries, one has $u \in L^{\frac{2n}{n-2}}(\tilde{D}_R)$. By Hölder's inequality,

$$\left| \int_{\tilde{D}_R} q|u|^2 dx \right| \leq \|q\|_{L^p(\tilde{D}_R)}\|u\|_{L^{2p'}(\tilde{D}_R)}^2,$$

where $p' := \frac{p}{p-1}$.

Choose $p' < \frac{n}{n-2}$. Then $p > \frac{n}{2}$. If $p > \frac{n}{2}$ and $u \in H^1(D_R)$, then

$$\left| \int_{\tilde{D}_R} q|u|^2 dx \right| \leq c(R)\|u\|_1^2.$$

Moreover, since the embedding $H^1(\tilde{D}_R) \to L^{2p'}(\tilde{D}_R)$ is compact for $p' < \frac{n}{n-2}$, it follows from the above arguments, from assumptions (A.3) and (A.4), and from the result of [7] that

$$\left| \int_{D_R} q|u|^2 dx \right| \leq v\|u\|_{H^1(D_R)}^2 + C(v)\|u\|_{L^2(D_R)}^2$$

for any $v > 0$, however small.

Assumption (A.5) and [7] imply

$$\left| \int_S \sigma|u|^2 ds \right| \leq v\|u\|_{H^1(D_R)}^2 + C(v)\|u\|_{L^2(D_R)}^2$$

for any $v > 0$, however small.

Thus, Lemma A.13 is proved. □

It follows from the above lemma that the norm $[B_\gamma(u, u)]^{1/2}$ is equivalent to $H^1(D')$ norm, the form $B_\gamma[\cdot, \cdot]$ is closed, symmetric, and densely defined in the Hilbert space $H := L^2(D')$. Therefore, this form defines a unique self-adjoint non-negative operator L in $H = L^2(D')$ with domain dense in $H^1(D')$. The points z with $\Im z \neq 0$ do not belong to its spectrum. Thus, problem (A.23), with $\varepsilon > 0$ has a unique solution in $H^1(D')$. We have proved the following.

Proposition A.14   *Problem (A.23) has a unique solution $w_\varepsilon \in H^1(D')$.*

In the proof of this proposition we have not used the compactness of the support of the potential $q(x)$ and gave the proof valid for $q(x)$ satisfying (A.3).

A.3.2   THE LIMITING ABSORPTION PRINCIPLE

In the above subsection we have proved that for each $\varepsilon > 0$, problem (A.23) admits a solution w_ε. In this section we shall prove that one can take the limit for ε going to zero, and get the solution of (A.20)–(A.22).

We first prove the following fundamental lemma.

Lemma A.15   *Suppose $\psi_n \in H^1_{loc}(D') \cap H^0_{-s}(D')$, with $s > 1$, and, in the weak sense,*

$$L\psi_n - \left(k^2 + i\varepsilon_n\right)\psi_n = h_n \tag{A.25}$$
$$\Gamma^R\psi_n = 0, \tag{A.26}$$

with $\varepsilon_n \downarrow 0$, $h_n \in L^2(D')$ have compact support and $h_n \to h$ in $L^2(D')$, where h has compact support. Moreover, suppose

$$\|\psi_n\|_{0,-s} \leq M, \tag{A.27}$$

where $M > 0$ is a constant independent of n. Then there exists a subsequence of $\{\psi_n\}_{n\in N}$, denoted again by $\{\psi_n\}_{n\in N}$, and a $\psi \in H^2_{\text{loc}}(D') \cap H^1_{-s}(D')$, with $s > 1$, such that:

$$\psi_n \longrightarrow \psi \qquad \text{strongly in} \quad H^2_{\text{loc}}(D'), \tag{A.28}$$
$$\psi_n \longrightarrow \psi \qquad \text{strongly in} \quad H^0_{-s}(D'), \tag{A.29}$$

with ψ solving the limiting problem:

$$L\psi - k^2\psi = h, \tag{A.30}$$
$$\Gamma^R\psi = 0. \tag{A.31}$$

Moreover,

$$\lim_{r\to\infty} \int_{|x|=r} |\psi_r - ik\psi|^2 ds = 0. \tag{A.32}$$

Proof. *Step 1: Convergence of ψ_n in $L^2_{\text{loc}}(D')$ and weak convergence in $H^1(D_r) \forall r > R$.*

Due to (A.27), there exists a subsequence $\psi_n \in H^0_{-s}(D')$ which converges weakly in $L^2_{loc}(D')$:

$$\psi_n \longrightarrow \psi \quad \text{weakly in} \quad L^2_{\text{loc}}(D'). \tag{A.33}$$

Let us prove that this sub-sequence ψ_n is bounded in $H^1_{loc}(D')$ uniformly with respect to n. If this is proved, then one gets:

$$\psi_n \longrightarrow \psi \quad \text{weakly in} \quad H^1_{\text{loc}}(D'), \psi_n \longrightarrow \psi \quad \text{strongly in} \quad L^2(D_r) \quad \forall r > R, \tag{A.34}$$

for a sub-sequence, denoted also by ψ_n.

To prove that

$$\|\psi_n\|_{H^1(D_r)} < c(r), \quad \forall r > R,$$

take in the weak formulation of (A.25), similar to (A.23), the test function $\varphi := \psi_n\beta(x)$, where $\beta(x)$ is a cut-off function, $0 \leq \beta(x) \leq 1$, $\beta(x) = 1$ for $|x| < r$, $\beta(x) = 0$ for $|x| > r + 1$, $r > R$. It then follows from the analog of (A.23) (see (A.35) below) that $\|\psi_n\|_{H^1(D_r)} < c(r) \forall r > R$, where $c(r) > 0$ is a constant independent of n, as claimed.

Similarly, one can prove that, if $s > 1$, then

$$\|\psi_n - \psi\|_{H^1_{-s}(D')} \to 0, \quad n \to \infty.$$

This is done as follows: write (A.23) with $w = \psi_n$ and subtract from (A.23) with $w = \psi_m$. Use (A.1), (A.30), (A.34), (A.38)–(A.40) (see below) to get the desired conclusion.

If ψ_n is a weak solution of the problem (A.25)–(A.26), one can write:

$$\int_{D'} \left[a_{ij} \partial_i \psi_n \partial_j \bar\varphi + (q - k^2 - i\varepsilon_n) \psi_n \bar\varphi \right] dx + \int_S \sigma \psi_n \bar\varphi ds = (h_n, \varphi) \tag{A.35}$$
$$\forall \varphi \in \tilde{H}^1(D').$$

Using (A.33) and (A.34) one can pass to the limit in the above equation and get:

$$\int_{D'} \left[a_{ij} \partial_i \psi \partial_j \bar\varphi + (q - k^2) \psi \bar\varphi \right] dx + \int_S \sigma \psi \bar\varphi ds = (h, \varphi) \tag{A.36}$$
$$\forall \varphi \in \tilde{H}^1(D') \ .$$

Step 2: Convergence of ψ_n in $H^2_{\mathrm{loc}}(D')$.

By the elliptic regularity, one concludes that ψ_n and ψ are in $H^2_{\mathrm{loc}}(D')$. Moreover, due to the density of $\tilde{H}^1(D')$ in $L^2(D')$, it follows from (A.35) and (A.36) that ψ_n and ψ satisfy (almost everywhere in D') (A.25) and (A.30), respectively.

Subtract (A.30) from (A.25), and use the elliptic estimate to get:

$$\|\psi_n - \psi\|_{H^2(D_1)} \le c \left(\|\psi_n - \psi\|_{L^2(D_2)} + \|h_n - h\|_{L^2(D_2)} + \varepsilon_n \|\psi_n\|_{L^2(D_2)} \right) \tag{A.37}$$
$$\forall D_1 \subset\subset D_2 \subset\subset D' ,$$

where $c > 0$ is a constant which depends on D_1 and D_2 but not on n. This estimate proves (A.28), since $\varepsilon_n \to 0$ and $\|\psi_n\|_{L^2(D_2)}$ is bounded.

Step 3: The representation formula and the radiation condition.

If ψ_n satisfies (A.25), if assumption (A.2) holds and if $q(x) = 0$ for $|x| > R$, then the following representation formula holds:

$$\psi_n(x) = \int_{S_R} \left[\psi_n(s) \partial_N g_{\varepsilon_n}(x, s) - g_{\varepsilon_n}(x, s) \partial_N \psi_n(s) \right] ds \qquad \text{for} \quad x \in B'_R, \tag{A.38}$$

where $\mathrm{Im}\sqrt{k^2 + i\varepsilon_n} > 0$ and

$$g_{\varepsilon_n} = \frac{e^{i\sqrt{k^2 + i\varepsilon_n}|x-y|}}{4\pi|x-y|}$$

is the Green's function of the operator $\Delta + (k^2 + i\varepsilon_n)$ and N in (A.38) and below stands for the normal to S_R pointing into B'_R. One can pass to the limit in (A.38) (due to the convergence of ψ_n in $H^2_{\mathrm{loc}}(D')$ proved in Step 2), and get the following representation formula for ψ:

$$\psi(x) = \int_{S_R} \left[\psi(s) \partial_N g(x, s) - g(x, s) \partial_N \psi(s) \right] ds, \tag{A.39}$$

where

$$g = \frac{e^{ik|x-y|}}{4\pi|x-y|}$$

is the Green function of the operator $\Delta + k^2$ in R^3.

Equation (A.39) implies that ψ satisfies the radiation condition (A.32).

Step 4: A uniform estimate of the behavior of ψ_n at infinity.

The representation formulas (A.38) for ψ_n and (A.39) for ψ, imply the following uniform estimate for the behavior of ψ_n at infinity (in R^3):

$$\sup_n (|\psi_n| + |\nabla \psi_n|) \leq \frac{c}{|x|} \qquad \text{for} \quad x \in B'_R, \tag{A.40}$$

where $c > 0$ is a constant independent of $x \in B'_R$. This estimate will be crucial in the next step. Note that (A.34), (A.40), and the estimate $\|\psi_n\|_{H^1(D_r)} \leq c(r) \ \forall r \geq R$ imply that $\psi \in H^1_{-s}(D')$, $s > 1$.

Step 5: Convergence of ψ_n in $H^0_{-s}(D')$ and the conclusion of the proof.

To complete the proof of the Lemma we have to prove the convergence property (A.29). Estimate (A.40) and the assumption $s > 1$ imply:

$$
\begin{aligned}
\|\psi_n - \psi\|^2_{0,-s} &= \int_{D'} \frac{|\psi_n - \psi|^2}{(1 + |x|^2)^{s/2}} dx \\
&= \int_{B_R \cap D'} \frac{|\psi_n - \psi|^2}{(1 + |x|^2)^{s/2}} dx + \int_{D'_R} \frac{|\psi_n - \psi|^2}{(1 + |x|^2)^{s/2}} dx \\
&\leq \sup_{B_R \cap D'} \left[\frac{1}{(1 + |x|^2)^{s/2}} \right] \|\psi_n - \psi\|^2_{L^2(B_R \cap D')} + 4\pi c^2 \int_R^\infty \frac{1}{r^2} \frac{1}{(1 + r^2)^{s/2}} r^2 dr \\
&\leq \eta,
\end{aligned}
\tag{A.41}
$$

with $\eta > 0$ arbitrarily small.

In the last step of the above estimate we have chosen R so that $4\pi c^2 \int_R^\infty dr/(1 + r^2)^{s/2} \leq \eta/2$, and (A.33) has been used. This proves (A.29) in R^3. In other space dimensions the proof is analogous. □

With the help of Lemma A.15 we prove the following *a priori* estimate for the solutions w_ε of the problem (A.23). In estimate (A.42) we take $0 < \varepsilon < 1$, but we could take $0 < \varepsilon < \varepsilon_0$, where $\varepsilon_0 > 0$ is an arbitrary small fixed number.

Proposition A.16   *The solution w_ε of problem (A.23) satisfies the following a priori estimate:*

$$\sup_{0 < \varepsilon < 1} \|w_\varepsilon\|_{0,-s} \leq c. \tag{A.42}$$

Proof. We prove this Proposition by contradiction. Suppose (A.42) is false. Find $\varepsilon_n \downarrow 0$ such that

$$\|w_{\varepsilon_n}\|_{0,-s} \geq n \ .$$

Define $\psi_n := w_{\varepsilon_n}/\|w_{\varepsilon_n}\|_{0,-s}$. Clearly, ψ_n satisfy all the hypotheses of Lemma A.15 with $h_n = f/\|w_{\varepsilon_n}\|_{0,-s}, h_n \to 0$ in $L^2(D')$. Lemma A.15 yields that $\psi_n \to \psi$ where ψ solves the problem:

$$L\psi - k^2\psi = 0,$$
$$\Gamma^R\psi = 0,$$
$$\lim_{r\to\infty}\int_{|x|=r}|\psi_r - ik\psi|^2 ds = 0.$$

By the uniqueness Theorem A.12 we get $\psi \equiv 0$.

This contradicts to the fact that $\|\psi_n - \psi\|_{0,-s} \to 0$ and $\|\psi_n\|_{0,-s} = 1$. Proposition A.16 is proved. $\qquad\square$

If w_ε satisfies (A.42) one can apply the Lemma A.15 to $\psi_n = w_{\varepsilon_n}$ for some $\varepsilon_n \downarrow 0$, and with $h_n = f$. We have therefore proved the existence of a w solving the problem (A.20)–(A.22) and the main result of this section.

Theorem A.17 *Scattering problem (A.8)–(A.10) with a compactly supported potential q has a unique solution u of the form $u = w + \zeta u_0$. Here $w \in H^2_{loc}(D') \cap H^1_{-s}(D')$, with $s > 1$, and ζ is a function defined above formula (A.20).*

A.4 EXISTENCE OF THE SCATTERING SOLUTION FOR DECAYING POTENTIALS

In this section we prove that the scattering problem (A.8)–(A.10), with a potential $q \in H^1_s$ has a solution. We shall prove the existence of the solution in the same way we did for the scattering problem with a compactly supported potential. First, we look for a solution u of the form $u = w + \zeta u_0$ so that the problem is reduced to (A.20)–(A.22). Then, we prove the existence of a unique solution w_ε for scattering problem (A.23). Finally, we shall prove that one can take the limit $\varepsilon \downarrow 0$ and get the solution of (A.20)–(A.22).

A.4.1 EXISTENCE FOR THE EQUATION WITH ABSORPTION

Proposition A.18 *Problem (A.23), with $\varepsilon > 0$, has a unique solution $w_\varepsilon \in H^1(D')$.*

The proof of this proposition is the same as the one of Proposition A.14.

A.4.2 THE LIMITING ABSORPTION PRINCIPLE

In this subsection we prove that the solution w_ε of problem (A.23) converges, as $\varepsilon \downarrow 0$, to the solution of (A.20)–(A.22).

The main step is the proof of Lemma A.21 below.

We first state two Lemmas.

Lemma A.19   *Suppose that* $|g(x, y)| \leq c|x - y|^{-1}$, *and* $|f(x)| \leq c\left(1 + |x|^2\right)^{-s/2}$, $x, y \in D'$ *and* $s > 3$. *Then*

$$\left| \int_{B'_R} g(x, y) f(y) dy \right| + \left| \int_{B'_R} \nabla_x g(x, y) f(y) dy \right| \leq \frac{c}{1 + |x|} , \quad x \in D'. \qquad (A.43)$$

Proof. Let $r = |y|$, $\rho = |x|$, $y = r (\sin \theta \cos \phi, \sin \theta \sin \phi, \cos \theta)$, $u = \cos \theta$, and let r, θ and ϕ be the spherical coordinates of y.

One has

$$
\begin{aligned}
\left| \int_{B'_R} g(x, y) f(y) dy \right| &\leq \int_{B'_R \cap \{y : |y| \leq \rho/2\}} \frac{|f(y)|}{|x - y|} dy + \int_{\{y : |y| \geq \rho/2\}} \frac{|f(y)|}{|x - y|} dy \\
&\leq \frac{c}{|x|} \int_{B'_R} |f(y)| dy + c \int_{\{y : |y| \geq \rho/2\}} \frac{dy}{|x - y| (1 + |y|^2)^{s/2}} \\
&\leq \frac{c}{|x|} + 2\pi c \int_{\rho/2}^{\infty} dr \frac{r^2}{(1 + r^2)^{s/2}} \int_{-1}^{1} \frac{du}{(\rho^2 + r^2 - 2r\rho u)^{1/2}} \\
&\leq \frac{c}{|x|} + c_1 \int_{\rho/2}^{\infty} \frac{dr}{(1 + r^2)^{\frac{s-2}{2}} \min(r, \rho)} \leq \frac{c}{|x|}.
\end{aligned}
$$

The second integral in (A.43) can be estimated similarly.

If $s > 2$ in Lemma A.19, then the argument above yields $o(1)$ as $|x| \to \infty$, in place of $\frac{c}{|x|}$ term. By a similar argument one can prove that if $s > 2$ and $g = \frac{e^{ik|x-y|}}{4\pi|x-y|}$, then the function

$$h := \int_{B'_R} g(x, y) f(y) dy$$

satisfies the radiation condition:

$$|x|| \frac{\partial h}{\partial |x|} - ikh| \to 0, \quad |x| \to \infty$$

uniformly in the directions of x. This remark does not allow one to replace $s > 3$ in the assumption (A.3) by $s > 2$. The reason is: if q is not compactly supported, the function f defined by (A.20) contains the term $q(x)\zeta(x)u_0$ which decays as $O((1 + |x|^2)^{-\frac{s}{2}})$ for large $|x|$. If $2 < s < 3$, then the argument given in Lemma A.19 is not sufficient for getting estimate (A.43). It is probable that the basic result, Theorem A.23, can be established for $s > 2$ in (A.3), but some additional argument is needed for a proof of such a result. $\square$

Lemma A.20   *Suppose that $|g(x, y)| \leq c|x - y|^{-1}$ and $|f(x)| \leq c\left(1 + |x|^2\right)^{-s/2}$, $x, y \in D'$, and $\psi \in H^0_{-s}(D')$ with $s > 3$. Then*

$$\left|\int_{B'_R} g(x, y) f(y) \psi(y) dy\right| + \left|\int_{B'_R} \nabla_x g(x, y) f(y) \psi(y) dy\right| \leq \frac{c}{|x|}, \quad x \in D'. \quad (A.44)$$

Proof. Denote $|x| := \rho$, $|y| := r$, $T(x, y) := |x - y|^{-2}(1 + |y|^2)^{-s/2}$. One has:

$$\left|\int_{B'_R} g(x, y) f(y) \psi(y) dy\right|$$

$$\leq \left[\int_{B'_R} |g(x, y) f(y)|^2 \left(1 + |y|^2\right)^{s/2} dy\right]^{1/2} \left[\int_{B'_R} |\psi(y)|^2 \left(1 + |y|^2\right)^{-s/2} dy\right]^{1/2}$$

$$\leq c\|\psi\|_{0,-s} \left[\int_{B'_R} T(x, y) \, dy\right]^{1/2}$$

$$\leq c_1 \left\{\left[\int_{B'_R \cap \{y \,:\, |y| \leq \rho/2\}} T(x, y) \, dy\right]^{1/2} + \left[\int_{\{y \,:\, |y| \geq \rho/2\}} T(x, y) \, dy\right]^{1/2}\right\}$$

$$\leq \frac{c}{|x|} + 2\pi c \left[\int_{\rho/2}^{\infty} dr \frac{r^2}{(1 + r^2)^{s/2}} \int_{-1}^{1} \frac{du}{\rho^2 + r^2 - 2r\rho u}\right]^{1/2} \leq \frac{c}{|x|}.$$

The second integral in (A.44) can be estimated similarly.   □

We can now prove a lemma analogous to Lemma A.15 of the previous section.

Lemma A.21   *Suppose $\psi_n \in H^1(D') \cap H^0_{-s}(D')$, with $s > 3$, and, in the weak sense,*

$$L\psi_n - \left(k^2 + i\varepsilon_n\right) \psi_n = h_n, \quad (A.45)$$

$$\Gamma^R \psi_n = 0, \quad (A.46)$$

where $\varepsilon_n \downarrow 0$, $h_n \in L^2(D')$, $|h_n(x)| \leq c(1 + |x|^2)^{-s/2}$, $s > 3$, and $h_n \to h$ in $L^2(D')$, where $|h(x)| \leq c(1 + |x|^2)^{-s/2}$. Moreover, suppose

$$\|\psi_n\|_{0,-s} \leq M, \quad s > 3, \quad (A.47)$$

where M is a constant independent of n. Then there exists a subsequence of $\{\psi_n\}_{n \in N}$, denoted again by $\{\psi_n\}_{n \in N}$, and a $\psi \in H^2_{\text{loc}}(D') \cap H^1_{-s}(D')$, such that:

$$\psi_n \longrightarrow \psi \quad \text{in} \quad H^2_{\text{loc}}(D'), \quad (A.48)$$

$$\psi_n \longrightarrow \psi \quad \text{in} \quad H^0_{-s}(D'), \quad (A.49)$$

where ψ solves the following problem:

$$L\psi - k^2\psi = h, \tag{A.50}$$
$$\Gamma^R\psi = 0, \tag{A.51}$$

$$\lim_{r\to\infty} \int_{|x|=r} |\psi_r - ik\psi|^2 ds = 0. \tag{A.52}$$

Proof. *Step 1: Convergence of ψ_n in $L^2_{\mathrm{loc}}(D')$ and weak convergence of ψ_n in $H^1(D_r)$, $r \geq R$.*
If (A.47) holds, then

$$\psi_n \longrightarrow \psi \qquad \text{strongly in} \qquad L^2_{\mathrm{loc}}(D'), \tag{A.53}$$
$$\psi_n \longrightarrow \psi \qquad \text{weakly in} \qquad H^1(D_r) \quad \forall r \geq R. \tag{A.54}$$

This is proved in Lemma A.15.

Step 2: Convergence ψ_n in $H^2_{\mathrm{loc}}(D')$.
One has

$$\psi_n \longrightarrow \psi \quad \text{strongly in} \quad H^2_{\mathrm{loc}}(D'). \tag{A.55}$$

This is proved as in Lemma A.15.

As in the Section A.3, ψ_n satisfies (A.45) and (A.46), and ψ satisfies (A.50) and (A.51).

Step 3: The representation formula for ψ_n and the radiation condition.
If ψ_n satisfies (A.45) then the following representation formula holds for $x \in B'_R$:

$$\psi_n(x) = \int_{S_R} [\psi_n(s)\partial_N g_{\varepsilon_n}(x,s) - \partial_N\psi_n(s)g_{\varepsilon_n}(x,s)]\, ds$$
$$+ \int_{B'_R} g_{\varepsilon_n}(x,y)h_n(y)dy - \int_{B'_R} g_{\varepsilon_n}(x,y)q(y)\psi_n(y)dy, \tag{A.56}$$

where N in the above formula and below denotes the normal to S_R pointing into B'_R. Since ψ_n converges in $H^2_{\mathrm{loc}}(D')$ one can pass to the limit in (A.56) and get:

$$\psi(x) = \int_{S_R} [\psi(s)\partial_N g(x,s) - \partial_N\psi(s)g(x,s)]\, ds$$
$$+ \int_{B'_R} g(x,y)h(y)dy - \int_{B'_R} g(x,y)q(y)\psi(y)dy. \tag{A.57}$$

Equation (A.57) implies that ψ satisfies the radiation condition.

Step 4: A uniform estimate of the behavior of ψ_n at infinity.

As in the case of a compactly supported potential one gets the following estimate for the behavior of ψ_n at infinity:

$$\sup_n \left(|\psi_n| + |\nabla \psi_n|\right) \leq \frac{c}{|x|} \qquad \text{for} \quad x \in B'_R, \tag{A.58}$$

where $c > 0$ is a constant independent of $x \in B'_R$. This follows from (A.56) and (A.57).

The additional terms which appear because the potential and the source term are not compactly supported, are estimated in Lemmas A.19 and A.20. Estimates (A.54) and (A.58) imply $\psi \in H^1_{-s}(D')$, $s > 1$.

Step 5: Convergence of ψ_n in $H^0_{-s}(D')$ and the conclusion of the proof.

One proves that $\|\psi_n - \psi\|_{0,-s} \to 0$ as in Lemma A.15. This concludes the proof of Lemma A.21. □

Now one gets the following *a priori* estimate for the solution w_ε of the problem (A.23).

Proposition A.22   *The solution of problem (A.23) satisfies the following a priori estimate:*

$$\sup_{0 < \varepsilon < 1} \|w_\varepsilon\|_{0,-s} \leq c, \quad s > 1. \tag{A.59}$$

The proof is based on Lemma A.21, and is the same as the proof of Proposition A.16.

Let us state *the main result of this appendix*, whose proof is based on the *a priori* estimate (A.59) and Lemma A.21.

Theorem A.23   *Assume that conditions (A.1)–(A.5) hold and $\sigma(s)$ is a real-valued $L^\infty(S)$- function. Then the scattering problem (A.8)–(A.10) has a weak solution u of the form $u = w + \zeta u_0$, with $w \in H^2_{\text{loc}}(D') \cap H^1_{-s}(D')$, $s > 1$, w satisfies (A.23) and (A.22), ζ is defined above formula (A.20), and this solution is unique in the above space.*

Bibliography

[1] S. Agmon, Spectral properties of Schrödinger operators and scattering theory, *Annali della Scuola Normale Superiore di Pisa*, 4, N2, pp. 151–218, 1975. 33

[2] H. Federer, *Geometric Measure Theory*, Springer Verlag, Berlin, 1969. DOI: 10.1007/978-3-642-62010-2. 33

[3] T. Kato, Growth properties of solutions of the reduced wave equation with a variable coefficient, *Communications on Pure and Applied Mathematics*, 12, pp. 403–425, 1959. DOI: 10.1002/cpa.3160120302. 36

[4] V. Mazya, *Sobolev Spaces*, Springer Verlag, New York, 1985. DOI: 10.1007/978-3-662-09922-3. 32, 33

[5] A. G. Ramm, *Scattering by Obstacles*, Reidel, Dordrecht, 1986. DOI: 10.1007/978-94-009-4544-9. 1, 2, 3, 6, 8, 9, 10, 17, 22, 23, 36

[6] A. G. Ramm, Uniqueness theorems for inverse obstacle scattering problem in Lipschitz domains, *Applications Analysis*, 59, pp. 377–383, 1995. DOI: 10.1080/00036819508840411. 31

[7] A. G. Ramm, A necessary and sufficient condition for compactness of embedding, *Vestnik Leningrad University ser. Mathem., Mech., Astron.*, N1, pp. 150–151, 1963. 39, 40

[8] A. G. Ramm, Investigation of the scattering problem in some domains with infinite boundaries I, II, *Vestnik Leningrad University ser. Mathem., Mech., Astron.*, N7, pp. 45–66; N19, pp. 67–76, 1963. 12

[9] A. G. Ramm, Scattering by obstacles and potentials, *World Scientific Publishing*, Singapore, 2017. DOI: 10.1142/10473. xiii, xiv, xv, 1, 2, 6, 8, 9, 13, 16, 17, 22, 24, 26

[10] A. G. Ramm, Creating materials with a desired refraction coefficient, *IOP Concise Physics*, Morgan & Claypool Publishers, San Rafael, CA, 2017. DOI: 10.1088/978-1-6817-4708-8. xiv, 2, 17

[11] A. G. Ramm, A uniqueness theorem in scattering theory, *Physics Review Letters*, 52, N1, p. 13, 1984. DOI: 10.1103/physrevlett.52.13. xiv

[12] A. G. Ramm, Uniqueness of the solution to inverse obstacle scattering problem, *Physics Letters A*, 347, N4–6, pp. 157–159, 2005. DOI: 10.1016/j.physleta.2005.08.088. xiv

[13] A. G. Ramm, Inverse scattering with under-determined scattering data, *Mathematical Modelling of Natural Phenomena, (MMNP)*, 9, N5, pp. 244–253, 2014. xiv

[14] A. G. Ramm, Uniqueness of the solution to inverse obstacle scattering with non-over-determined data, *Applied Mathematics Letters*, 58, pp. 81–86, 2016. DOI: 10.1016/j.aml.2016.02.006. xiv, 2, 17

[15] A. G. Ramm, A numerical method for solving 3D inverse scattering problem with non-over-determined data, *Journal of Pure and Applied Mathematics*, 1, N1, pp. 1–3, 2017. DOI: 10.1016/0893-9659(88)90155-3. xv

[16] A. G. Ramm, Uniqueness of the solution to inverse scattering problem with backscattering data, *Eurasian Mathematical Journal, (EMJ)*, 1, N3, pp. 97–111, 2010. DOI: 10.1063/1.3666985. xv

[17] A. G. Ramm, Uniqueness of the solution to inverse scattering problem with scattering data at a fixed direction of the incident wave, *Journal of Mathematical Physics*, 52, 123506, 2011. DOI: 10.1063/1.3666985. xv

[18] A. G. Ramm, *Inverse Problems*, Springer, New York, 2005. xv, 29

[19] A. G. Ramm, Uniqueness theorem for inverse scattering problem with non-over-determined data, *Journal of Physics A, (FTC)*, 43, 112001, 2010. DOI: 10.1088/1751-8113/43/11/112001. xv

[20] A. G. Ramm, Scattering of acoustic and electromagnetic waves by small bodies of arbitrary shapes, *Applications to Creating New Engineered Materials*, Momentum Press, New York, 2013. DOI: 10.5643/9781606506226. 8

[21] A. G. Ramm, A uniqueness theorem for inverse scattering problem with non-over-determined data, *Engineering Science Letters, (ESL)*, 2018:3, pp. 1–5, 2018. DOI: 10.1088/1751-8113/43/11/112001. 17

[22] A. G. Ramm, Inverse obstacle scattering with non-over-determined data, *Global Journal of Mathematical Analysis, (GJMA)*, 6(1), pp. 2–6, 2018. DOI: 10.1109/mmet.2016.7544097. xiv, 17

[23] A. G. Ramm and Cong Tuan Son Van, A numerical algorithm for solving 3D inverse scattering problem with non-over determined data, *Journal of Applied Mathematics and Statistics Applied*, 2, N1, pp. 11–13, 2018. xv

[24] A. G. Ramm and A. Ruiz, Existence and uniqueness of scattering solutions in non-smooth domains, *Journal of Mathematical Analysis and Applications*, 201, pp. 329–338, 1996. DOI: 10.1006/jmaa.1996.0258. 31

[25] A. G. Ramm and M. Sammartino, Existence and uniqueness of the scattering solutions in the exterior of rough domains, in the book *Operator Theory and Its Applications, American Mathematical Society*, Fields Institute Communications, vol. 25, pp.457–472, Providence, RI, 2000. 31

[26] A. G. Ramm, Inverse scattering with non-over-determined data, *Journal of Advances in Mathematics*, 16, pp. 1–4, 2019. DOI: 10.24297/jam.v16i0.8089. xiv, 2, 17

[27] G. Verchota, Layer potentials and regularity for the Dirichlet problem in Lipschitz domains, *Journal of Functional Analysis*, 59, pp. 572–611, 1984. DOI: 10.1016/0022-1236(84)90066-1. 31

[28] A. Volpert and S. Hudjaev, *Analysis in Classes of Discontinuous Functions and Equations of Mathematical Physics*, Martinus, Dordrecht, 1985. 39

[29] T. Wolff, A property of measures and an application to unique continuation, *Geometric and Functional Analysis*, 2, N2, pp. 225–284, 1992. DOI: 10.1007/bf01896975. 36

Author's Biography

ALEXANDER G. RAMM

Alexander G. Ramm, Ph.D., was born in Russia, immigrated to the U.S. in 1979, and is a U.S. citizen. He is Professor of Mathematics with broad interests in analysis, scattering theory, inverse problems, theoretical physics, engineering, signal estimation, tomography, theoretical numerical analysis, and applied mathematics. He is an author of 690 research papers, 16 monographs, and an editor of 3 books. He has lectured in many universities throughout the world, presented approximately 150 invited and plenary talks at various conferences, and has supervised 11 Ph.D. students. He was a Fulbright Research Professor in Israel and in Ukraine, distinguished visiting professor in Mexico and Egypt, Mercator professor, invited plenary speaker at the 7th PACOM, won the Khwarizmi international award, and received other honors. Recently, he solved inverse scattering problems with non-over-determined data and the many-body wave-scattering problem when the scatterers are small particles of an arbitrary shape; Dr. Ramm used this theory to give a recipe for creating materials with a desired refraction coefficient. He gave a solution to the refined Pompeiu problem, and proved the refined Schiffer's conjecture.

Printed in the United States
by Baker & Taylor Publisher Services